D0065607

Vibrational Spectroscopies for Adsorbed Species

Alexis T. Bell, EDITOR

University of California, Berkeley

Michael L. Hair, EDITOR

Xerox Research Center of Canada

Based on a symposium

sponsored by the Division of

Colloid and Surface Chemistry

at the 178th Meeting of the

American Chemical Society,

Washington, D.C.,

September 12–13, 1979.

A C S   S Y M P O S I U M   S E R I E S 137

AMERICAN CHEMICAL SOCIETY

WASHINGTON, D. C.   1980

Library of Congress CIP Data
Vibrational spectroscopies for adsorbed species.
(ACS symposium series; 137 ISSN 0097–6156)
Includes bibliographies and index.
1. Vibrational spectra—Congresses. 2. Infrared
spectrometry—Congresses. 3. Surface chemistry—Congresses.
I. Bell, Alexis T., 1942– . II. Hair, Michael L.,
1934– . III. American Chemical Society. Division
of Colloid and Surface Chemistry. IV. Title: Adsorbed
species. V. Series: American Chemical Society. ACS
symposium series; 137.

QC454.V5V534   541.3'453   80-21181
ISBN 0-8412-0585-X ACSMC8 137 1–295 1980

Copyright © 1980

American Chemical Society

All Rights Reserved. The appearance of the code at the bottom of the first page of each
article in this volume indicates the copyright owner's consent that reprographic copies of
the article may be made for personal or internal use or for the personal or internal use of
specific clients. This consent is given on the condition, however, that the copier pay the
stated per copy fee through the Copyright Clearance Center, Inc. for copying beyond that
permitted by Sections 107 or 108 of the U.S. Copyright Law. This consent does not extend
to copying or transmission by any means—graphic or electronic—for any other purpose,
such as for general distribution, for advertising or promotional purposes, for creating new
collective works, for resale, or for information storage and retrieval systems.

The citation of trade names and/or names of manufacturers in this publication is not to be
construed as an endorsement or as approval by ACS of the commercial products or services
referenced herein; nor should the mere reference herein to any drawing, specification,
chemical process, or other data be regarded as a license or as a conveyance of any right or
permission, to the holder, reader, or any other person or corporation, to manufacture, repro-
duce, use, or sell any patented invention or copyrighted work that may in any way be
related thereto.

PRINTED IN THE UNITED STATES OF AMERICA

QC
454
V5 V534

ACS Symposium Series

M. Joan Comstock, *Series Editor*

Advisory Board

David L. Allara

Kenneth B. Bischoff

Donald G. Crosby

Donald D. Dollberg

Robert E. Feeney

Jack Halpern

Brian M. Harney

Robert A. Hofstader

W. Jeffrey Howe

James D. Idol, Jr.

James P. Lodge

Leon Petrakis

F. Sherwood Rowland

Alan C. Sartorelli

Raymond B. Seymour

Gunter Zweig

468537

FOREWORD

The ACS SYMPOSIUM SERIES was founded in 1974 to provide a medium for publishing symposia quickly in book form. The format of the Series parallels that of the continuing ADVANCES IN CHEMISTRY SERIES except that in order to save time the papers are not typeset but are reproduced as they are submitted by the authors in camera-ready form. Papers are reviewed under the supervision of the Editors with the assistance of the Series Advisory Board and are selected to maintain the integrity of the symposia; however, verbatim reproductions of previously published papers are not accepted. Both reviews and reports of research are acceptable since symposia may embrace both types of presentation.

CONTENTS

PREFACE

The identification of species adsorbed on surfaces has preoccupied chemists and physicists for many years. Of all the techniques used to determine the structure of molecules, interpretation of the vibrational spectrum probably occupies first place. This is also true for adsorbed molecules, and identification of the vibrational modes of chemisorbed and physisorbed species has contributed greatly to our understanding of both the underlying surface and the adsorbed molecules. The most common method for determining the vibrational modes of a molecule is by direct observation of adsorptions in the infrared region of the spectrum. Surface spectroscopy is no exception and by far the largest number of publications in the literature refer to the infrared spectroscopy of adsorbed molecules. Up to this time, the main approach has been the use of conventional transmission IR and work in this area up to 1967 has been summarized in three books. The first chapter in this volume, by Hair, presents a necessarily brief overview of this work with emphasis upon some of the developments that have occurred since 1967.

One of the major advances in the past decade has been the maturation of the electronic revolution. This has had its effect on surface spectroscopy, with regard to instrumentation for transmission IR, but particularly for sensitivity gains that have made reflectance techniques the preferred alternative for fundamental studies. In the transmission mode, the commercial development of the Fourier transform IR spectrometer has led to significant advantages in the determination of the vibrational spectra of adsorbed species. This is covered in the chapter by Bell.

In the transmission mode, IR spectroscopy is limited by sensitivity requirements to the study of surfaces exhibiting relatively high surface area, i.e., porous oxides and supported metals. A major advance in the past decade has been achieved by using reflection techniques. Modulation of the incident beam or ellipsometry usually are employed and optimization of these techniques now enables the IR spectra of highly absorbing molecules to be obtained on single crystal surfaces. As a result, it has become possible to compare vibrational data with data obtained from other surface diagnostic techniques (i.e., LEED and AES). Chapters describing the determination of these reflection absorption spectra are given by Allara, Pritchard, and Dignam.

Surface wave spectroscopy provides yet another means for obtaining absorption spectra of species present on a single crystal surface. In this case, IR radiation is coupled to the sample in such a way that it propagates laterally along the surface for a finite distance before it is decoupled and detected. Use of this technique permits a larger absorption to be achieved than normally is obtained by means of reflection techniques. The theory and practice of surface wave spectroscopy are discussed by Bell.

Raman spectroscopy is a complementary technique that yields the same vibrational information as is obtained in conventional IR absorption spectroscopy. Here, however, the exciting radiation is in the visible region of the spectrum. Application of this technique to surfaces has been difficult because of practical problems associated with fluorescence. This problem now has been essentially resolved, and recent progress in the application of Raman spectroscopy to surfaces is reviewed by Morrow.

Conventionally, in IR spectroscopy the measured absorptions are caused by the fundamental vibrations of the atoms within a molecule. In many cases, important information can be obtained by examining the overtones of these fundamental vibrations. This can be done in the transmission mode, but elegant work using diffuse reflectance spectroscopy is discussed by Klier.

The final chapters of this book review the progress of three recently developed techniques that provide information about the vibrational states of adsorbed molecules. Perhaps the most important of these techniques is electron energy loss spectroscopy that, despite its inherent low resolution, gives valuable information on vibrational modes that are either inactive in the IR, or inaccessible because of experimental difficulties. The applications of this technique are discussed in two chapters by Somorjai and Weinberg. The review of new experimental techniques concludes with presentations on inelastic electron tunneling spectroscopy and neutron scattering by Kirtley and Taub.

In assembling this collection of review/progress articles, the editors have tried to provide an update of all techniques used to determine the vibrational structure of molecules adsorbed on surfaces. The symposium itself provided a forum whereby the leading workers in the field could interact and rationalize their various approaches. No single technique will ever give all the information required to describe an adsorbed molecule and the recent developments relating the vibrational spectroscopies and the UHV techniques are, thus, particularly exciting. Further overlap between these techniques will undoubtedly occur and inevitably lead to a deeper understanding of the molcular structure of adsorbed species.

Department of Chemical Engineering   ALEXIS T. BELL
Berkeley, California

Xerox Research Center of Canada   MICHAEL L. HAIR
Mississauga, Ontario, Canada

May 19, 1980

Transmission Infrared Spectroscopy for High Surface Area Oxides

MICHAEL L. HAIR

Xerox Research Centre of Canada, 2480 Dunwin Drive,
Mississauga, Ontario, L5L 1J9, Canada

Interest in the vibrational spectra of adsorbed molecules is at least 40 years old. The past ten years have seen the development of many novel techniques for determining the vibrational spectra of adsorbed species and this symposium brings together a state-of-the-art survey of these techniques. In one's ethusiasm for the recent advances made in any subject there is a tendency to forget the parent technique and its steady contribution to our knowledge. In this case, the parent is simple transmission infrared spectroscopy. This paper, therefore, is an attempt to briefly present an overview of some of the developments which have occurred in the application of transmission infrared spectroscopy to surface studies with emphasis upon results generated in the past 10 years. For more detailed information on work published prior to 1967 the reader is referred to three texts which have appeared on this subject (1-3).

Because of the maturity of the method the advances in technique are incremental rather than revolutionary. Perhaps the major new developments have been in the instrumental area where the ready availability of the Fourier Transform instruments has led to its introduction to surface studies. The ease of obtaining spectra and the advantages associated with the direct computation of data will be discussed in a separate paper (4).

Transmission IR still remains the best method for examining insulating oxide surfaces and over the past decade there has developed a considerable understanding of many surfaces, particularly those of silica, alumina, molecular sieves and complex catalysts. The objective of this paper, therefore, will be to demonstrate how some of the recent advances have been made. Clearly it is not possible to discuss all the materials studied by transmission IR and the author has chosen to use the surface properties of silica to illustrate the type of understanding that is now available.

Some History

The first application of transmission infrared spectroscopy to the study of adsorbed species appears to be the work of Buswell *et al* in 1938 (5). Those authors pressed a montmorillonite clay into a disc which was then "dried" at various temperatures. The spectra they obtained bear a remarkable similarity to many others that have been produced in the literature over the next forty years: the authors were clearly able to resolve bands due to hydroxyl groups associated with the clay lattice and to adsorbed water which was slowly removed as a function of drying.

0-8412-0585-X/80/47-137-001$05.00/0
© 1980 American Chemical Society

Almost twenty years elapsed before the next advances. In the 1950's the Russian School (Terenin, Yaroslavski, Kiselev) (6) studied the structure of porous glass and made the first assignments for hydroxyl groups on such surfaces; the Cambridge School (Shephard, Little, Yates) (7) investigated the process of physical adsorbtion and the rotational motion of physcially adsorbed molecules; and the American workers (Eischens, Francis, Pliskin) (8) applied the technique to supported metal catalysts and initiated the application of this technique to the study of catalysts. The work of Eischens and co-workers on CO/Ni is remarkable in that the spectra obtained (and the interpretation) have stood the test of time and are confirmed most elegantly by some of the studies on single crystal surfaces which will be presented in later chapters.

Experimental

In the application of transmission infrared spectroscopy to the study of surfaces it is important that high surface area materials be used in order that the resultant spectrum contains a considerable contribution from the surface as distinct from the bulk of the sample. Typically, these samples have been pressed into thin self-supporting discs which are then inserted in the path of the infrared radiation in the spectrometer. Developments in this area in the past decade have been aimed mainly at quantifying the infrared data and two basic types of vacuum cell have emerged: those in which the sample is moved in and out of the beam into a furnace above the spectrometer, and those in which the furnace is built around a static sample holder which is permanently held in the beam. The latter is clearly a more desirable system but suffers from the experimental disadvantages that the furnace must be constructed within the confines of the infrared spectrometer, thus giving rise to problems associated with the cooling of the infrared transmitting windows. An excellent cell capable of temperatures up to 600°C under UHV conditions has been described (9).

Silica

A classical paper on the adsorption of water on silica surfaces appeared from the General Electric Laboratories in 1958 (10). One set of spectra from this paper are redrawn in Fig. 1. The results clearly demonstrate the difference between the two forms of high surface area silica commonly found in the laboratory: the silica which is precipitated from solution and is widely used as a dessicant and the finely divided silica (Cabosil, Aerosil, etc.) which is prepared by flame oxidation of $SiCl_4$ at elevated temperatures. Experimentally, MacDonald made pressed discs of each of the forms of silica, placed them in a simple IR cell in the beam of a spectrometer as described earlier and recorded the spectra. The solid black lines (a) are the spectra recorded at room temperature and the difference between the two silicas is readily apparent. The Cabosil spectrum shows distinct structure with bands being readily observed at 3747, 3660 and 3520 cm^{-1}. In the case of the precipitated silica there is complete absorption between 3750 and 3000 cm^{-1}. On evacuating the samples at room temperature changes are seen in the spectra which can be related to the removal of physically adsorbed water from the surface. This is done by comparing (a) and (b). A broad band (d) centered around 3400 cm^{-1} has been removed by the evacuation and this is coincident with the removal (not shown) of a band at 1625 cm^{-1}. These are clearly due to the stretching and bending

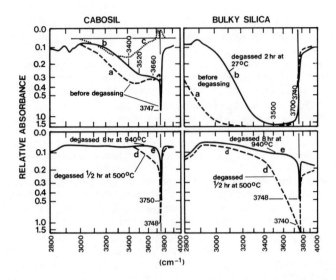

Journal of Physical Chemistry

Figure 1.   The IR spectra of Cab–O–Sil and bulky silica: (a) before degassing; (b) after degassing for 3 h at 30°C in vacuo; (c) difference between (a) and (b); (d) degassed 30 min at 500°C in vacuo; (e) degassed 8.5 h at 940°C in vacuo (10)

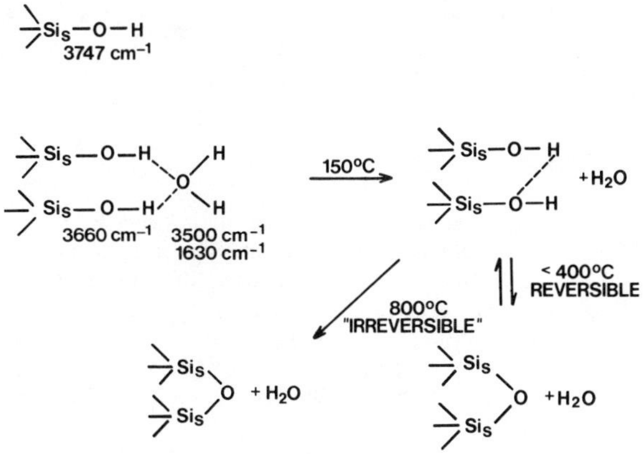

Figure 2.   Schematic of hydration–dehydration on a silica surface

motions of molecular water. On heating the sample to higher temperatures further changes can be observed. In both samples the band at 3660 cm^{-1} becomes more distinct as the heating temperature is raised until it eventually is seen as a shoulder on the 3747 cm^{-1} band. After the sample has been degassed at 940°C in vacuum, both the precipitated silica and the Cabosil exhibit identical spectra which consist of one very narrow, sharp band at 3747 cm^{-1}.

One important observation which can be drawn from these spectra is the fact that, at least at low coverages, the removal of molecular water from the surface of the silica does not particularly affect the intensity of the band at 3747 cm^{-1}. Thus, it can be concluded that the water is not specifically interacting with this group during the adsorption at low coverages and is therefore sitting on other parts of the surface. (This view is widely held, but definitive evidence is not available. Interaction certainly occurs at higher partial pressure but this would be expected in a random rather than a specific adsorption process.)

The interpretation of the bands seen in these spectra is now almost universally accepted although there may be some disagreement about detail and fine structure. Thus, the band at 3747 cm^{-1} is assigned to a hydroxl group attached to a silicon atom on the surface. The OH group is situated so that it is vibrating freely and is unperturbed by its neighbours. The band at 3660 cm^{-1} is attributed to surface hydroxyl groups which are sufficiently close together that they are hydrogen bonded to each other and are thus perturbed from the 3747 cm^{-1} position. It is usually accepted that the frequency shift is related to the strength of the H-bonding interaction, the larger shifts corresponding to stronger interaction and it is noted that dehydration proceeds from the lower frequency end. Other absorptions in this region are attributed to molecular water adsorbed on the surface. The dehydration which occurs as the temperature of the sample is raised clearly proceeds by a mechanism in which the H-bonded SiOH groups are removed from the surface, the more strongly H-bonded groups being removed initially.

Rehydration of the surface is found to be reversible only if the pretreatement temperature is kept below about 400°C. Above that temperature a restructuring of the surface apparently occurs. The removal of the H-bonded groups removes the adsorption site for water thus giving rise to a surface which does not adsorb water very readily at low pressure (Fig. 2).

Sample Preparation

One of the major experimental problems in studying surfaces by the transmission techniques has always been the sample preparation. Indeed, the quality of the spectra was limited by the quality of the pressed disc and good transmission below 1300 cm^{-1} was not possible. This precluded observation of the region where many bending modes would be expected and prevented complete assignment of surface structures. After much experience it is now possible to prepare very thin, high quality discs of Cabosil and other oxides (10 mg/cm^2) and the resultant increase in available information is illustrated by the work of Morrow and his co-workers on Cabosil (11-14). Thus, Morrow and Cody were able to obtain the spectra shown in Fig. 3 in the region between 800 and 1000 cm^{-1}. The effect of heating the sample is clearly shown and can be compared with the data recorded 20 years earlier (Fig. 1). As the sample is heated the low frequency tail in the 3660 cm^{-1} region is removed from the spectrum until a single, sharp band is observed at 3747 cm^{-1}. At the same time, an increase in transmission is noted between 800 and 1000 cm $^{-1}$ and new structure appears at temperatures in excess of 500°C. By re-examining the adsorption of water, ammonia, and other molecules on these surfaces Morrow et al were able to positively identify these bands as being due to a new surface siloxane group formed during the dehydroxylation (cf Fig. 2). This new group will react with water to reform surface silanol groups *but not at the 3747 cm*$^{-1}$ position. Clearly, these groups are contributing to asymmetry previously observed in the 3747 cm^{-1} band and could account for small discrepancies observed, for instance, in kinetic data.

Isotopic Exchange

The use of isotopic exchange in all forms of infrared spectroscopy is well known and has been applied to the identification of surface species in recent years. The work, of Morrow and his co-workers in this respect is of interest. (14)

It has been known for many years that the surface hydroxyl groups can be readily converted to OD simply by repeated exposure to D_2O. Early experiments with this type of exchange were used in order to manipulate vibrational modes to portions of the spectrum which were more transparent. Thus, for instance, isotopic substitution of D for H in surface silanol groups moves the fundamental stretching frequency from 3747 to 2750 cm^{-1}. Peri (15) used this technique before computer subtraction was so readily available in order to determine the bond angle of the surface Si_s-OH group. By recording sample background in the 2800 cm^{-1} region and then isotopically shifting the Si_sOH vibration he was able to identify P and R branches of the spectrum superimposed upon the previous background. This enabled calculation of the Si_s-O-H bond angle.

A second interesting application of the D_2O exchange has been used by Hambelton, Hockey and Taylor (16) to emphasize the care that must be exercised when interpreting the spectra obtained with self-supporting discs. By preparing self-supporting discs of Aerosil under varying applied pressures they were able to demonstrate that the effect of increased pressure was to cause perturbation of some of the freely vibrating hydroxyl groups in such a manner that they no longer became available for D_2O exchange and thus could no longer be considered as surface species.

In addition to isotopic substitution of D for H in the surface silanol groups, Morrow (13) has also demonstrated that up to 60% of the surface silanol grups may be substituted with O^{18}. This substitution can be exceptionally valuable in assigning bands involving the oxygen atom. Thus, when BF_3 is adsorbed on silica the spectrum shown in Fig. 4 (A) is observed. The two bands at 1450 and 1500 cm^{-1} were both originally assigned to BF vibrations. Repetition of the experiment on an O^{18} exchanged silica surface, however, gives the spectrum shown in Fig. 4 (B). The fact that the band previously observed at 1500 cm^{-1} now appears as a doublet clearly indicates that that peak must be

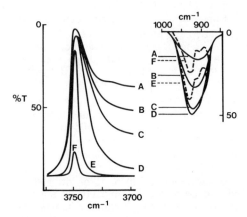

Journal of Physical Chemistry

Figure 3.   Background spectra of untreated silica (10 mg cm⁻²) after degassing under vacuum: (A) 180°C; (B) 300°C; (C) 460°C; (D) 675°C; (E) 900°C; (F) 1200°C (14)

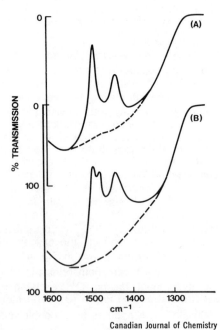

Figure 4.   Spectrum obtained after adsorption of $^{10}BF_3$ on dehydrated Cab–O–Sil: (A) untreated silica; (B) ^{18}O-exchanged silica; (– – –) indicates the background spectrum of Cab–O–Sil before adsorption of BF_3 (13)

Canadian Journal of Chemistry

due to a vibrational mode involving oxygen whereas the band at 1450 cm^{-1} is a pure BF vibration.

Quantitative Aspects of Adsorption

I.   Adsorption Isotherms

In discussing the hydration/dehydration of silica in Fig. 1 it was pointed out that at low surface coverages it appeared that the water molecules preferentially adsorbed on the H-bonded hydroxyl groups rather than on the freely vibrating surface hydroxl group. In the case of molecules containing lone pair electrons this is not the case and the IR method clearly demonstrates that such molecules can form a strong hydrogen bond with the surface silanol group. This is shown in Fig. 5. In this example (17) a silica surface has been dehydrated so as to contain both free OH groups and H-bonded OH groups. Diethylamine was then adsorbed on the surface and it can be seen that whereas the H-bonded groups are little affected by the adsorption of the diethylamine, the freely vibrating group is perturbed, in this case by almost l000 cm^{-1}. There is also a considerable increase in intensity of the perturbed band. Such experiments can be used to quantitatively determine the adsorption isotherm of a molecule on the freely vibrating surface hydroxyl group (18). Thus, measuring the decrease in intensity of the 3747 cm^{-1} band as a function of gas pressure reveals the adsorption isotherm *on the freely vibrating group.* Such an isotherm obtained for the adsorption of diethylether on silica is shown in Fig. 6 (a). For comparative purposes the total volumetric adsorption isotherms obtained for diethylether on this silica is shown (b), as is the adsorption isotherm measured on a silica surface in which the surface silanol has been removed from the surface by chemical reaction.

It is to be noted that the adsorption isotherm on the freely vibrating group is of the simple Langmuir type and that there is excellent agreement between the volumetric and spectroscopic data showing that the volumetric adsorption isotherm which is commonly measured is indeed the summation of the adsorption on the hydroxyl group plus the adsorption on the remainder of the surface (19).

2.   Surface Acidity.

The ability to obtain isotherms on individual groups also enable the calculation of the isosteric heat of adsorbtion on that site (as distinct from the heats of adsorption averaged over the whole surface which are normally obtained from adsorption isotherms). Relationships between the OH frequency shifts and heats of adsorbtion have been obtained this way (18). A relationship between the frequency shift and the ionisation potential of the adsorbing molecules has been demonstrated by several authors (20-22) and a theoretical explanation based on the Mullikan-Puranik approach to H-bonding has been given by Low and Cusamano (23).

The question of the acidity of silica, alumina and silica-alumina surfaces has always been of great interest to catalytic scientists. Previously, transmision infrared spectroscopy, particularly of pyridine adsorption, has been used to distinguish the presence of Lewis and Bronsted acid sites on oxide surfaces (24). The frequency shift of the surface OH group during adsorption now

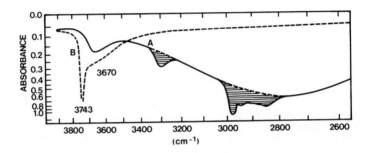

Journal of Chemical Physics

*Figure 5.   Spectrum of silica dried at 450°C: (A) diethylamine adsorbed on (B);
the cross-hatched areas are attributable to diethylamine (17).*

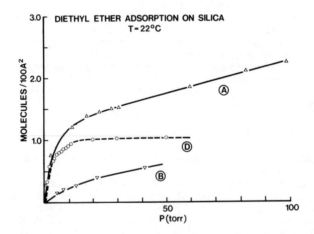

Journal of Physical Chemistry

*Figure 6.   Adsorption of diethyl ether on silica at 24°C: (A) volumetric isotherm
on silica containing free OH groups (+SiOSi); (B) volumetric isotherm on dehy-
droxylated silica; (D) spectroscopic isotherm on free OH groups (19)*

enables meaurements of the acidity of these groups to be estimated. The technique used is based upon the linear relationships which are observed when hydroxylic acids of varying strength are dissolved in solvents of varying basicity and is a well established technique in physical organic chemistry for the determination of acid dissociation constants. The linear relationship between frquency shift and pK_a has been the subject of numerous papers and the interested reader is referred to the work of Bellamy, Hallett and Williams (25). The use of frequency shifts to determine acid dissociation constants of surface groups was first demonstrated by Hair and Hertl (26) who obtained pK_a values for the hydroxyl groups on oxides of silica, magnesia, silica-alumina, phosphorus and boron. This work was later extended by Rouxhet and Sempels (27) who demonstrated that for adsorption of various bases on silica-alumina catalysts bands were observed that could be attributed to the perturbation of both the normal silanol group and an additional strong acid group. Strongly negative pK_a values (-7.0) were obtained and this is in agreement with the concept that silica-alumina cracking catalyst contain very strong Bronsted acid sites which are responsible for catalytic acitivty.

3.   Kinetics of Surface Reactions

Another area in which the quantitative application of infrared spectroscopy to silanol surfaces has proved useful has been in the study of the kinetics of the reactions of silicon tetrachloride and the other Group IV halides and the substituted chloromethylsilanes with the silica surface (28,29). Although not of prime interest to catalytic chemists these data are useful in that they provide considerable insight into the methods of surface deactiviation employed, for instance, in chromatography or as coupling agents in the attachment of other active species (i.e. enzymes) to silica surfaces. The results demonstrate that the silanising molecules react essentially only with the free surface silanol group. The reactions are kinetically straight forward but the mixed order reaction, which is in agreement with other analytical data, demonstrates that the detailed structure of the free surface OH group is not as simple as the IR spectra might indicate. With the other halides, direct chlorination of the surface is always observed and this is sometimes is the only raction. With BCl_3 the reaction is autocatalytic.

Conclusions

In the limited space available this paper has attempted to give an overview of the ways that transmission infrared spectroscopy has been applied to the study of high surface area materials. Developments in improved sample preparation and the use of isotopic substitution have been discussed. The more quantitative aspect of work accomplished in the last decade has been emphasized by giving examples of adsorbtion isotherms on individual sites and the subsequent reactivity of the adsorbed molecules with these sites.

No discussion has been devoted to the recent use of Fourier transform spectrometers rather than dispersion instruments. The ease with which the spectral data can be manipulated and background subtracted make the FT methods particularly useful for studies of surface species, particularly during catalytic reaction. Recently there has been a surge of interest in the coupling of computer subtraction techniques to conventional grating instruments. For many IR surface studies, where only limited frequency range is required, this may be a preferred approach. The reader is referred to recent work by Peri (30) who has developed such methods for the study of carbon monoxide on supported metal catalyst. Minor frequency shifts oberved in the CO stretching mode of the adsorbed gas have been related to subsequent catalytic activity.

No attempt has been made to discuss the voluminous literature on molecular sieves or supported metals and the reader is referred to papers by Nacchace and Uytterhoeven (31) and Sheppard (32) respectively. IR studies on adsorption from solution are particularly relevant to polymer adsorption and have been reviewed by Rochester (33).

LITERATURE CITED

1. Hair, M.L., "Infrared Spectroscopy in Surface Chemistry." Dekker, New York, 1967.

2. Little, L.H., "Infrared Spectra of Adsorbed Species." Academic Press, New York, 1966.

3. Kiselev, A.V., and Lygin, V.I., "Infrared Spectra of Surface Compounds." Wiley, New York, 1975.

4. Bell, A.T.   This volume.

5. Buswell, A.M., Krebs, K. and Rodebush, W.H., J. Am. Chem. Soc., 59, 2603 (1937).

6. Yaroslavskii, N.G. and Karyakin, A.V., Dokl. Akad. Nauk SSSR, 85, 1103 (1952).

7. Sheppard, N. and Yates, D.J.C., Proc. Roy Soc. (London), A238, 69 (1956).

8. - Eischens, R.P., Francis, S.A., and Pliskin, W.A., J. Phys. Chem., 60, 194 (1956).

9. Gallei, E. and Schadow E., Rev. Sci. Instr. 45, 1504 (1974).

10. McDonald, R.S., J. Phys. Chem., 62, 1168 (1958).

11. Morrow, B.A. and Cody, I.A., J. Chem. Soc., Faraday Trans. 1, 71, 102l (1975).

12. Morrow, B.A. and Devi, A., J. Chem. Soc., Faraday Trans. 1, 68, 403 (1972).

13. Morrow, B.A. and Devi, A., Can. J. Chem., 48, 2454 (1970).

14. Morrow, B.A. and Cody, I.A., J. Phys. Chem., 80, 1998 (1976).

15. Peri, J.B., J. Phys. Chem., 70, 2937 (1966).

16. Hambleton, F.H., Hockey, J.A. and Taylor, J.A.G., Nature 203, 138 (1965).

17. Basila, M.R., J. Chem. Phys., 35, 1151 (1961).

18. Hertl, W., and Hair, M.L., J. Phys. Chem. 72, 4676 (1968).

19. Hair, M.L. and Hertl, W., J. Phys. Chem. 73, 4269 (1969).

20. Basila, M.R., J. Chem. Phys. 35, 1151 (1961).

21. Kiselev, A.V., Surface Sci., 3, 292 (1965).

22. van Duijneveldt, F.B., Vermoortele, F.H., and Uytterhoeven, J.B., Discuss. Faraday Soc. 52, 66 (1971).

23. Cusamano, J.A. and Low, M.J.D., J. Catalysis 23, 214 (1971).

24. Parry, E.P., J. Catalysis, 2, 371 (1963).

25. Bellamy, L.J. and Williams, R.J., Proc. Roy. Soc. A 254, 119 (1960).

26. Hair, M.L. and Hertl, W., J. Phys. Chem. 74, 91 (1970).

27. Rouxhet, P.G. and Semples, R.E., J. Chem. Soc. Faraday Trans. 1 70, 2021 (1974).

28. Hair, M.L. and Hertl, W., J. Phys. Chem. 73, 2372 (1969).

29. Hair, M.L. and Hertl, W., J. Phys. Chem. 77, 17 (1973).

30. Peri, J.B., Preprints, Division of Petroleum Chemistry, Inc., A.C.S. Meeting, Miami Beach, 1978.

31. Nacchace, C., Taarit, Y.B. and Uytterhoeven, J.B., "Proceedings of the Symposium on Zeolites, Szeged, Hungary. (publ. Hungarian Academy of Sciences ed. H. Beyer).

32. Sheppard, N. and Nguyen, T.T., Adv. Infrared Raman Spectrosc. 5, 67 (1978).

33. Rochester, C.H., "Powder Technology 13, 157 (1976).

RECEIVED June 3, 1980.

Applications of Fourier Transform Infrared Spectroscopy to Studies of Adsorbed Species

ALEXIS T. BELL

Materials and Molecular Research Division, Lawrence Berkeley Laboratory and Department of Chemical Engineering, University of California, Berkeley, CA 94720

Infrared spectroscopy has been widely used to identify molecular structures present at the surface of catalysts and other solids (1-3). In most instances dispersive spectrometers, utilizing either a prism or a grating as the dispersing element, have proven to be satisfactory for such studies. The limitations of such instruments have been encountered, though, in certain instances. For example, when the radiation intensity emanating from the sample is very low, as is the case in working with strongly absorbing or scattering samples or in performing emission spectroscopy, wide slit settings and slow scanning times must be used to achieve a reasonable signal to noise ratio. Difficulties can also be encountered in observing complete spectra under dynamic conditions.

The introduction of commercial Fourier transform (FT) spectrometers in the early 1960's has made it possible, in part, to overcome the limitations associated with dispersive instruments and has helped to broaden the scope of problems amenable to investigation by infrared spectroscopy. The purpose of this review is to compare the performance of FT and dispersive spectrometers and to illustrate areas of application in which FT spectroscopy has proven advantageous for the study of adsorbed species. In view of these objectives only a limited treatment of the theory underlying FT spectroscopy will be presented here. More detailed discussions of this subject and examples of the applications of FT spectroscopy to other fields of chemistry can be found in references (4, 5, 6, 7, 8).

Optical Principles

To establish a quantitative basis for comparing the performance of dispersive and FT spectrometers, it is necessary to review first the optical principles governing the operation of each type of instrument. The primary purpose here will be to

0-8412-0585-X/80/47-137-013$05.75/0
© 1980 American Chemical Society

describe the means by which continuous radiation is resolved into its spectral components.

Dispersive Spectroscopy. The physics governing dispersive spectrometers will be illustrated in terms of the operation of a grating spectrometer. The optical layout for such an instrument is shown in Figure 1. Radiation emitted by the source is divided into two beams. One beam passes through the sample while the other serves as a reference. A rotating sectored mirror combines the two beams to form a single beam consisting of alternate pulses of radiation from the sample and reference beams. The combined beam enters the monochrometer through a slit, and is dispersed into its spectral elements by a grating. The radiation appearing at the exit slit of the monochrometer is determined by the diffraction relationship

$$n/\nu = 2g \cos\varepsilon \sin\theta \qquad\qquad (1)$$

where ν is the wavenumber for a given spectral element, n = 0,1,2... is the order of the diffraction, g is the spacing between rulings on the grating, ε is the angle between the incident and diffracted beams, and θ is the angle of rotation of the grating from the zero order (n=0). For any value of θ, equation 1 is satisfied by a number of values of ν corresponding to different values of n. Therefore, radiation of different wavenumbers corresponding to different orders of diffraction from the grating emerge from the exit slit. An optical filter is used to reject radiation from all but the desired order of diffraction.

After filtering, radiation from the monochrometer is focused onto a thermocouple detector. The alternating signal at the detector is amplified and fed to a servo-motor which moves a reference beam attenuator to equalize the intensity of the sample and reference beams. The alternating signal is thereby reduced, producing a state of equilibrium. The absorbance of a sample placed in the sample beam is determined by the extent of movement of the reference beam attenuator.

FT Spectroscopy. The principals governing the operation of an FT spectrometer can be understood best by following a beam of monochromatic radiation through the optical layout shown in Figure 2. The beam enters the source side of a Michelson interferometer where part of the beam is reflected to the fixed mirror by the beam splitter and part is transmitted to the moveable mirror. Following reflection, both beams recombine at the beam splitter. Since the path length of the beam reflected from the moveable mirror is in general slightly different from that of the beam reflected from the fixed mirror, the two beams interfere either constructively or destructively. Displacement of the moveable mirror at a fixed velocity, v, modulates the beam exiting from the interferometer at a frequency 2 vν. For a

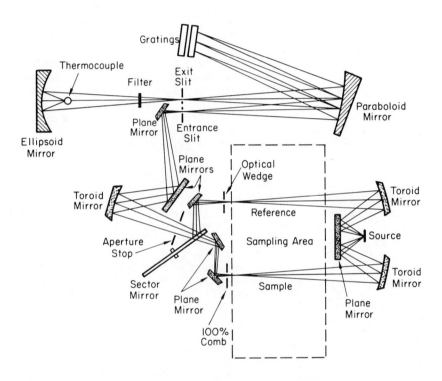

Figure 1.   Optical arrangement of a Perkin Elmer Model 457 grating spectrometer

mirror velocity of 0.2 cm/s, modulation frequencies between 1.6 KHz and 160 Hz are produced over the spectral range of 4000 to 400 cm^{-1}.

The modulated beam is directed through either the sample or reference side of the sample compartment and is finally focused on the detector. For most mid-infrared work, a triglycine sulfate (TGS) pyroelectric bolometer is used as the detector because of its very high frequency response (> 1 MHz).

The signal observed by the detector is measured as a function of the displacement or retardation, s, of the moveable mirror. For a monochromatic source, the interferogram, I(s), is a sinusoid exhibiting maxima for s = n/ν and minima for s = (n + 1/2)/ν, as shown in Figure 3. With an increase in the number of lines in the spectrum, the complexity of the interferogram increases so that it quickly becomes impractical to deduce any features of the spectrum solely by inspection of the interferogram.

To obtain a spectrum I(ν) from a measured interferogram, it is necessary to take the Fourier transform of the fluctuating portion of the interferogram, $\bar{I}(s)$ = [I(s) - 0.5 I(0)]. Thus,

$$I(\nu) = \int_{-\infty}^{\infty} \bar{I}(s) \cos(2\pi\nu s)\, ds \qquad (2)$$

An illustration of the relationship between $\bar{I}(s)$ and I(ν) for a spectrum consisting of three narrow lines is shown in Figure 4a.

In practice, the Fourier integral indicated in equation 2 cannot be executed over s from $-\infty$ to $+\infty$, since the interferogram can be determined experimentally only over a finite range ($-s_{max} \leq s \leq + s_{max}$). As a consequence, the integration can be performed only over a finite range. Figure 4b shows that truncation of the limits of integration results in a broadening of the spectral peaks.

An additional consequence of finite retardation is the appearance of secondary extrema or "wings" on either side of the primary features. The presence of these features is disadvantageous, especially when it is desired to observe a weak absorbance in proximity to a strong one. To diminish this problem the interferogram is usually multiplied by a triangular apodization function which forces the product to approach zero continuously for s = $\pm$ s$_{max}$. Fourier transformation of the apodized interferogram produces a spectrum such as that shown in Figure 4c.

Since the acquisition of an interferogram and its subsequent transformation to produce a spectrum requires a large amount of data handling and computation, these tasks are normally carried out with the aid of a dedicated minicomputer. The computer is also used to operate the spectrometer and to display spectra.

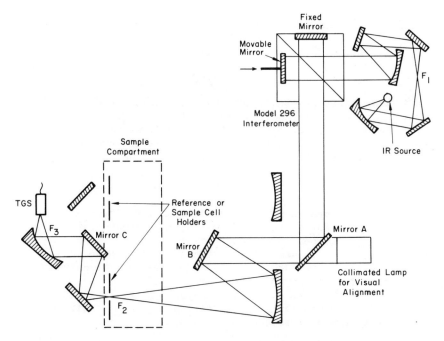

Figure 2. Optical arrangement of a Digilab Model FTS-14 FT spectrometer

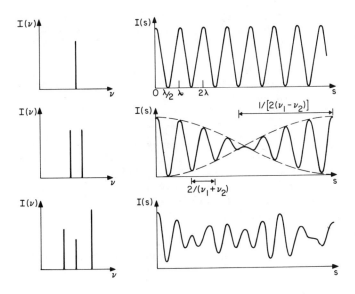

Figure 3. Spectra $I(\nu)$ and interferograms $\overline{I}(s)$ for one, two, and three narrow lines

Comparison of Fourier-Transform and Grating Spectrometers

Resolving Power.   The resolving power of a spectrometer is defined by the quantity

$$R = \nu/\Delta\nu \qquad (3)$$

and is determined by either the characteristics of the resolving element in the spectrometer or the finite dimensions of the radiation source as it appears at the resolving element. The resolving power of a grating spectrometer is constrained by the width of the monochrometer slits, w, and is given by

$$R_G = \frac{2f}{w} \tan\theta \qquad (4)$$

where f is the focal length of the collimating mirror in the monochrometer. For a fixed slit width, R_G decreases with increasing wavenumber (i.e., increasing θ) and as a consequence the resolution decreases. This pattern can be compensated by decreasing the slit width but only at the expense of decreasing the energy flux to the monochrometer.

In most instances the resolving power of an FT spectrometer is determined by the maximum retardation of the moveable mirror in the interferometer so that

$$R_I = \nu s_{max} \qquad (5)$$

From equations 3 and 5 it follows that $\Delta\nu = 1/s_{max}$. Consequently, the resolution of an FT spectrometer is fixed by the maximum retardation, but the resolving power increases with increasing wavenumber.

Single to Noise Ratio.   The signal to noise ratio for both interferometers and monochrometers can be determined in an identical fashion (9,10). The signal power, S, received at the detector is given by

$$S = B(T,\nu) \theta \eta \Delta\nu \qquad (6)$$

where $B(T,\nu)$ is the source brightness, which is a function of both of the source temperature and the wavenumber ν; θ is the optical throughput; η is the optical efficiency; and $\Delta\nu$ is the resolution. Assuming that the detector itself is the dominant noise source, the noise power, N, is then expressed as

$$N = NEP/t_m^{1/2} \qquad (7)$$

where NEP is the noise equivalent power of the detector and t_m is time for observation of a single spectral element of width $\Delta\nu$. Dividing equation 6 by equation 7 results in the following expression for the signal to noise ratio, S/N:

$$S/N = B(T,\nu) \ominus \eta \ \Delta\nu \ t_m^{1/2}/NEP \qquad (8)$$

Griffiths et al. (9) have compared the signal to noise ratios of interferometers and monochrometers under the assumption that the source temperature and resolution for both types of instrument are equivalent. Their analysis involves a comparison of the factors appearing in equation 8 and a representation of the advantage of an interferometer over a monochrometer as the ratio of the factors associated with each instrument. A summary of Griffith et al.'s (9) analysis is presented in the balance of this section.

The first advantage of interferometers over monochrometers is known as Fellgett's advantage. It results from the fact that an interferometer observes all spectral resolution elements simultaneously, whereas a monochrometer spends only a fraction of the total measurement time observing each resolution element. If the spectral range is ν_R, then the number of resolution elements, M, examined is $\nu_R/\Delta\nu$. Fellgett's advantage can be expressed in two ways. Spectra measured with equal signal to noise ratios on instruments with identical sources, detectors, optical throughput, and optical efficiency will theoretically be measured M times faster using an FT spectrometer than using a dispersive spectrometer. Conversely, spectra measured using an FT spectrometer will theoretically have a signal to noise ratio which is $M^{1/2}$ times higher than the corresponding spectra measured using a dispersive spectrometer in the same data acquisition time. A summary of both methods of expressing Fellgett's advantage is given in Table 1.

The second fundamental advantage of an FT spectrometer over a grating spectrometer is often called Jacquinot's advantage and derives from the increased optical throughput of an interferometer compared to a grating monochrometer.

The maximum optical throughput for a Michelson interferometer, Θ_{max}^I is given by

$$\Theta_{max}^I = 2\pi \ A^I \ \frac{\Delta\nu}{\nu_{max}} \qquad (9)$$

where A^I is the area of the interferometer mirrors and ν_{max} is the highest wavenumber in the spectrum. Since all spectral elements are incident on the detector at all times during the measurement, the throughput is the same for all wavenumbers. This fact puts the interferometer at a disadvantage compared with a dispersive spectrometer for which the gratings can be changed during measurement to allow increased throughput and efficiency for longer wavelength radiation.

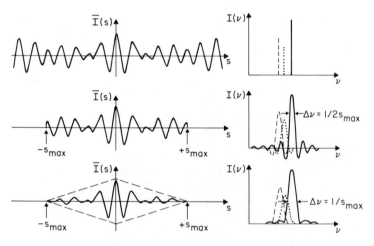

Figure 4.   Appearance of spectrum obtained by Fourier transformation of: (a) an infinite interferogram; (b) a finite interferogram; and (c) a finite interferogram with triangular apodization

Table I

Tabulated Values of Fellgett's Advantage

Resolution $\Delta \nu \, (cm^{-1})$	Range $\nu_R (cm^{-1})$	S/N Advan. M	t_m Advan. $M^{1/2}$
8	3600	450	21
2	3600	1800	42
0.5	3600	7200	85
0.5	400	800	28

The throughput of a grating spectrometer, Θ^G, is given by:

$$\Theta^G = \frac{hA^G}{fa\nu^2} \Delta\nu \qquad (10)$$

where h is the height of the entrance and exit slits, f is the focal length of the collimator, A^G is the area of the grating, and $a = g^{-1}$ is the spacing of the lines on the grating. The ratio of Θ^I to Θ^G gives the magnitude of Jacquinot's advantage:

$$\frac{\Theta^I}{\Theta^G} = 2\pi \frac{A^I fa\nu^2}{A^G h\nu_{max}} \qquad (11)$$

A plot of this ratio for two commercial spectrometers is shown in Figure 5. The discontinuities in this plot are associated with differences in the optical properties of the gratings used in the dispersive instrument to cover each of the spectral regions. It is observed further that while Jacquinot's advantage is very high at the high wavenumber end of each region, it falls off as ν^2 with decreasing wavenumber.

The multiplex and throughput advantages obtained with an FT spectrometer are partially compensated by the fact that the TGS detector used with such instruments has a considerably higher NEP than the thermocouple detector used with grating spectrometers. For example, for radiation chopped at 15 Hz, the NEP of thermocouple detectors can be as low as 1×10^{-10} W Hz$^{-1/2}$, whereas at the same frequency the NEP of a TGS detector is typically 2×10^{-9} W Hz$^{-1/2}$. As the modulation frequency of the signal increases the NEP of both detectors deteriorates. This effect is not significant for grating spectrometers since the signal is modulated by a rotating sector, and the modulation frequency of the beam is constant for all wavenumbers. By contrast, the modulation frequency for an interferometer is directly proportional to the wavenumber of the radiation and modulation frequencies lie typically in the range of 10^2 Hz to 1 KHz. An illustration of the NEP ratio for the two types of detector is shown in Figure 6.

The optical transmission efficiencies associated with interferometers and monochrometers are quite different, as seen in Figure 7. The efficiency of a Ge/KBr beam splitter designed for mid-infrared performance is nearly 90% between 4000 and 1000 cm^{-1} but falls off rapidly for smaller wavenumbers. By contrast, gratings exhibit high efficiency over much shorter wavenumber ranges. This characteristic is compensated, though, by changing gratings.

Finally one must also recognize that the duty cycle efficiencies of the two types of instrument are also different. The duty cycle efficiency, ζ, is defined as the ratio of the time

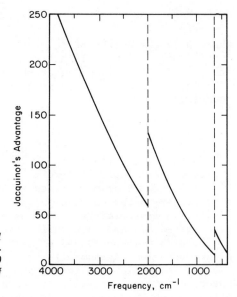

Figure 5.   The ratio of the calculated throughputs of the Digilab FTS-14 Fourier Spectrometer and the Beckman 4240 grating spectrometer as a function of wavenumber (9)

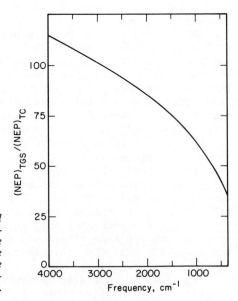

Figure 6.   The ratio of NEP of TGS and TC detector as a function of wavenumber (9). The chopping frequency for the TC detector is 15 Hz while that for the TGS detector is $2 \times 0.3164 \, v$ Hz, where 0.3164 cm s^{-1} is the velocity of the moving mirror on the Digilab spectrometer.

during which the spectrometer actively takes measurements to the total cycle time. For an interferometer, a portion of the total total measurement time is used to reset the moveable mirror. This results in a value of $\zeta^I \sim 0.9$ for single beam operation. Smaller values of ζ^I are obtained for interferometers operated in a double-beam mode, since the moveable mirror must be left stationary for a fraction of the cycle time to allow the detector to stabilize each time the beam is switched from the sample to the reference position. With an optical null grating spectrometer the chopper is used not only to modulate the beam but also to alternate the beam between sample and reference channels. Thus, it takes approximately the same time to measure a transmittance spectrum using a double beam optical null spectrometer as it takes to measure a single-beam spectrum with the same S/R. Hence, for this type of spectrometer ζ^G may be assigned a value of 2.

The overall advantages of an FT spectrometer over a grating spectrometer can be represented by either the ratio of S/N's for an equivalent t_m or the square root of the ratio of t_m's for an equivalent S/N. Figure 8 summarizes the advantages of an FT spectrometer over a grating spectrometer. Curve A represents the calculated advantage for spectra measured between 4000 and 400 cm^{-1} at a resolution of 2 cm^{-1} and equal data acquisition times. Curve B compares the performance of a Digilab FTS-14 spectrometer to that of a Beckman 4240 grating spectrometer. It is observed that the practical advantage of the FT spectrometer closely follows the predicted theoretical advantage, but is a factor of four smaller. Griffiths et al. (9) propose that this discrepancy might be ascribed to the manner of calculating Fellgett's advantage. Based upon results obtained by Tai and Harwit (11), Griffiths et al. (9) suggest that Fellgett's advantage should be $(M/8)^{1/2}$ instead of $M^{1/2}$. Application of this correction leads to curve C, which is in better agreement with the experimental results.

The results presented in Figure 8 clearly indicate that for the conditions of comparison chosen, an FT spectrometer can provide an advantage over a grating spectrometer in terms of higher signal to noise ratios and very much lower data acquisition times. It should be noted, however, that the magnitude of the advantage is diminished if either a narrower wavenumber range than 4000 to 400 cm^{-1} or a lower resolution than 2 cm^{-1} is considered. As a result, the basis of comparison must be clearly stated before a quantitative assessment can be made of the advantages of working with an FT spectrometer.

<u>Data Acquisition, Storage, and Display</u>. Fourier-transform spectrometers require access to an on-line computer in order to permit rapid transformation of interferograms to spectra, and, in fact, the development of commercial instruments was in large

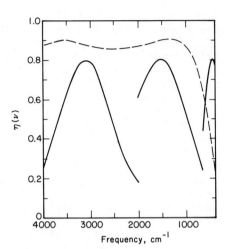

Figure 7. The efficiency of the gratings used on the Beckman 4240 spectrometer (——) and the calculated efficiency of a Ge/KBr beamsplitter whose performance is optimized for 2250 cm⁻¹ (– – –) (9)

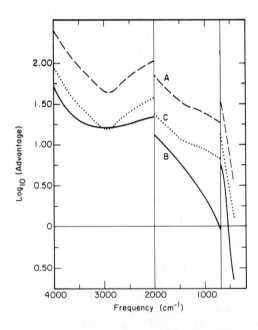

Figure 8. (A) Theoretical advantage of an FT spectrometer with the same parameters as a Digilab FTS-14 over a grating spectrometer with the same parameters as the Beckman 4240, both instruments operating at 2 cm⁻¹ resolution (9). (B) Measured advantage of a Digilab FTS-14 over a Beckman 4240, both operating at 2 cm⁻¹ resolution. (C) Theoretical advantage of an FT spectrometer over a grating spectrometer (same instrumental parameters as Curve A), assuming Fellgett's advantage is $(M/8)^{1/2}$ and not $M^{1/2}$.

part, made possible by the introduction of minicomputers in the mid-1960's. The availability of a computer provides a number of additional advantages including data storage on either magnetic tape or discs, comparison and subtraction of spectra, and great flexibility in displaying spectra. Moreover, the computer can be used to smooth data, to subtract baselines, and to compute integral absorbances. It should be noted, however, that many of the same data acquisition and handling functions can be achieved by interfacing dispersive spectrometers to a minicomputer or microprosessor, and that at present there are several commercial grating spectrometers which are fully integrated with such data-handling systems. As a result, the presence of a computer as a part of an FT spectrometer cannot be regarded as a feature providing such an instrument with an advantage over dispersive spectrometers.

Applications of Fourier-Transform Spectroscopy

Transmission. The most frequently used technique for observing the structure of species present at the surface of a solid is direct transmission of an infrared beam through a disc formed by pressing a fine powder of the solid. Difficulties arise when the solid strongly adsorbs or scatters the incident radiation. If a grating spectrometer is used to measure the spectrum, these difficulties are diminished partially by working with thin samples, opening the monochromator slits, and scanning very slowly. These approaches are successful particularly for samples which do not change composition with time. However, when there is a need to observe many samples or the sample composition changes over the period of observation, then the long periods of data acquisition needed for dispersive spectrometers becomes disadvantageous. Under these circumstances the shorter observation times resulting from Fellgett's advantage make the use of an FT spectrometer attractive.

Low and coworkers (12-17) have utilized FT spectroscopy extensively to characterize species adsorbed on the surface of strongly absorbing and scattering solids such as CaO and MgO. Their work has shown that spectra exhibiting good resolution can be recorded in 5 to 10 min. To obtain similar spectra using a grating spectrometer would require one to three hours.

Bouwman and Freriks (18,19) have also noted the advantages of FT spectroscopy and have used this technique to study the adsorption of CO on a silica-supported nickel catalyst at temperatures between 70 and 180°C. These authors point out that FT spectroscopy is particularly advantageous for in situ observation of heated samples since the radiation emitted by the sample is not modulated by the interferometer and hence does not contribute to the fluctuating portion of the interferogram. While this conclusion is correct, it should be noted that radiation emitted by a sample can be excluded in a dispersive instrument by chopping the source beam prior to its incidence on the sample and using a phase-sensitive lock-in amplifier to observe only the ac portion of the detected radiation.

   Specular and Diffuse Reflectance. Reflectance techniques have been applied primarily to samples which do not permit observation by transmission. Flat surfaces such as those of metal foils and single crystals can be studied by specular reflectance, whereas rough surfaces such as those of powders must be observed by diffuse reflectance. In both cases FT spectroscopy offers strong advantages in terms of the time required to acquire a spectrum.

   Most applications of FT spectroscopy to the study of adsorbed species by means of specular reflection have been limited to observations of CO adsorbed on metal surfaces. In a short communication, Low and McManus [20] have reported detection of a band at 2090 cm^{-1} for CO chemisorbed on a platinum foil using a multiple reflection cell and a low resolution interferometer. A more detailed study of the adsorption of CO on palladium has been reported by Harkness [21] and Apse [22]. The optical arrangement used for these studies is shown in Figure 9. The reflection cell consisted of many layers of palladium foil arranged so that multiple reflections could be achieved. Using this arrangement Harkness [21] detected a band at about 2060 cm^{-1} attributed to the C-O stretching vibration of chemisorbed CO. In a subsequent effort, Apse [22] noted the presence of a broad, weak band at 375 cm^{-1}, which was assigned to the Pd-C stretching vibration.

   Very recently, Bailey and Richards [23] have shown that a high degree of sensitivity for adsorbed species can be achieved by measuring the absorption of infrared radiation on a thin sample cooled to liquid helium temperature. The optical arrangement used in these studies is shown in Figure 10. The modulated beam produced by the interferometer is introduced into the UHV sample chamber and reflected off a thin slice of monocrystalline alumina covered on one side by a 1000 Å film of nickel or copper. Radiation absorbed by the sample is detected by a doped germanium resistance thermometer. The minimum absorbed power detected by this device when operated at liquid helium temperature is 5×10^{-14} W for a 1 Hz band width. With this sensitivity absorbtivities of 10^{-4} could be measured.

   Spectra of CO adsorbed on nickel and copper films obtained by Bailey and Richards [23] are shown in Figure 11. Carbon monoxide adsorption at 77°K resulted in the appearance of only a single band at 2097 cm^{-1} for nickel and 2109 cm^{-1} for copper. Further CO exposure of the sample at 1.5°K resulted in the observation of a second band near 2143 cm^{-1} on both metals. The low frequency band in each case was identified with chemisorbed CO while the high frequency band was associated with physisorbed CO. Bailey and Richards [23] have also used their apparatus to obtain spectra of N_2 chemisorbed on nickel, and benzene and

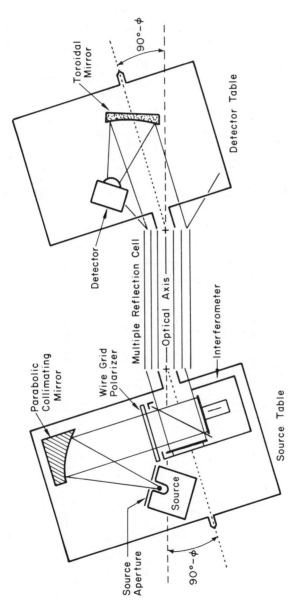

Figure 9. Optical arrangement used by Harkness (21) for specular reflectance spectroscopy

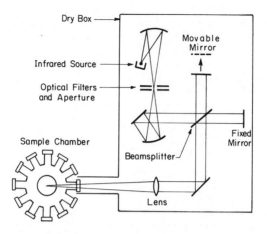

Figure 10.   Optical arrangement used by Bailey and Richards (23) for specular
reflectance spectroscopy

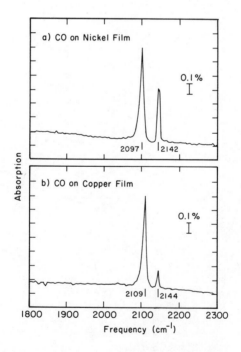

Figure 11.   Spectra of CO adsorbed on Cu and Ni films (23)

methane physisorbed on copper.

When the reflecting surface is rough the incident radiation is reflected diffusely. To record a spectrum in this case the reflected radiation must be collected and focused on a detector. Several devices for performing diffuse reflectance spectroscopy have been described in the literature (24, 25). Figure 12 illustrates a very effective optical arrangement recently reported by Fuller and Griffith (25). The collimated beam from an interferometer is reflected off a paraboloidal mirror which focuses the beam onto a sample positioned at one focus of an ellipsoidal mirror. Because of this geometry, any specularly reflected radiation from the sample is directed back through the hole in the ellipsoidal mirror. The diffusely reflected component is collected and then focused on a detector.

Diffuse reflectance spectroscopy has been widely used to characterize the surface of solids as well as films and coatings present on a solid substrate (6). The application of this technique to the study of adsorbed species has been much more limited (26) and, thus far, has not involved the use of FT spectroscopy.

Emission Spectroscopy. Emission spectroscopy offers an alternative to reflectance spectroscopy as a means for recording the spectrum of species present at the surface of opaque solids. Measurement of emission spectra is complicated, however, by the fact that the radiation intensity of the emitting sample is low, making it difficult to achieve a good signal to noise ratio. The utilization of an FT spectrometer is advantageous, in this case, since it allows the superposition of a large number of spectra, thereby permitting an improvement to be made in the signal to noise ratio (6, 27).

Low and coworkers (28, 29, 30) have pioneered the application of FT spectrometers to the observation of emission spectra from solid surfaces. An example taken from a study by Low and Coleman (29) is illustrated in Figure 13. Spectrum A shows the emission of a clean aluminum foil at 33°C. Spectrum B is of the same surface after it had been wiped with an oleic acid-coated paper and then wiped dry. Spectrum C is the same as B, but taken with twice the number of scans and twice the gain. The bands near 1700, 1650, and 1600 cm^{-1} are attributed to the carbonyl-stretching frequency of undissociated acid, to the asymmetric O-C-O vibrations of the oleate ion, and to the symmetric O-C-O vibrations, respectively. Similar bands were detected at 100°C, as shown in spectrum D, indicating that the oleic acid had reacted with the aluminum surface to form an oleate.

A further example of the use of FT spectroscopy for the observation of emitted radiation is provided by the work of Kember and Sheppard (31). By observing the radiation emitted by a copper surface heated in air, these authors were able to detect

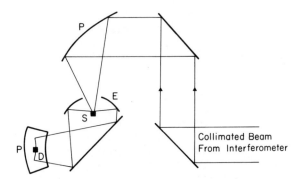

Figure 12. Optical arrangement by Fuller and Griffiths (25) for diffuse reflectance spectroscopy: P, paraboloidal mirror; E, ellipsoidal mirror; S, sample; D, detector

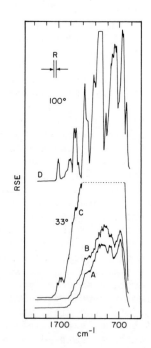

Figure 13. Emission spectra of oleic acid on aluminum: A, emission of the clean foil, 120 scans, 33°C; B, after oleic acid addition, 120 scans, 33°C; C, repetition of spectrum B, 240 scans, at twice the gain; D, oleic-acid-treated foil at 100°C, 120 scans

the formation of oxide films. Both Cu_2O and CuO were observed when the copper was heated above 300°C as evidenced by the appearance of emission bands at 620 cm^{-1}(Cu_2O) and 500 cm^{-1} (CuO).

Time Resolution. Time-resolved studies of surface species are of considerable interest in the field of catalysis since they offer a means for investigating the kinetics of adsorption and surface reaction and for distinguishing between species active and inactive in catalysis (32, 33, 34). Dispersive spectrometers can be used for this purpose (33, 35) but are restricted to the observation of either a single frequency or a narrow range of frequencies, unless the dynamics of the observed phenomenon are very slow compared to the time required for the acquisition of a spectrum. FT spectroscopy allows these limitations to be surmounted and opens up the possibility of recording complete spectra very rapidly.

If the time constants associated with the processes being observed are large compared to the time for the acquisition of a single interferogram (0.1 to 1 sec), then time-resolved spectra can be obtained in a sequential fashion. Savatsky and Bell (36) have recently utilized this approach to study the adsorption of NO onto a silica-supported rhodium catalyst. Figure 14 illustrates the manner in which the intensities of the bands observed at 1680, 1830, and 1910 cm^{-1} change with time of exposure. The The band at 1680 cm^{-1}, which is assigned to N-O vibrations of $NO_a^{\delta-}$, grows to a maximum intensity and then diminishes for exposures above 100 sec. During this latter period, the band at 1910 cm^{-1}, which is assigned to $NO_a^{\delta+}$, intensifies and, after exposure times of 800 sec or more, becomes the dominant spectral feature. Since the band at 1630 cm^{-1} is characteristic of NO adsorbed on a reduced rhodium surface and the band at 1910 cm^{-1} is characteristic of NO adsorbed on an oxidized surface, the relative intensities of these bands provide an indication of the extent to which the rhodium surface, is oxidized by NO. The band observed at 1830 cm^{-1} is ascribable to NO adsorbed in a neutral state and appears to be relatively insensitive to the progressive oxidation of the catalyst.

Time-resolved spectra can also be obtained for processes characterized by time constants which are smaller than the time to acquire an interferogram. To make this possible the experiment must be repeated many times in a reproducible fashion. The data for an interferogram are acquired in the following manner. The moveable mirror of the interferometer is advanced stepwise through M position increments. At each retardation, s_j, the mirror is held stationary for one cycle of the experiment and N+1 measurements of the interferometric signal are made. These measurements are represented by $I(t_o,s_j)$ through $I(t_N,s_j)$. Figure 15 illustrates the distribution of measurements obtained using this scheme. By assembling the signals recorded for a

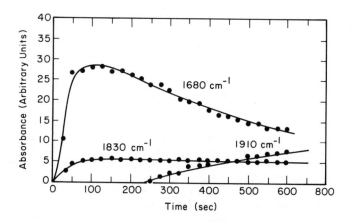

*Figure 14. Time evolution of bands associated with NO adsorption on 4% Rh/
SiO₂ catalyst; adsorption temperature, 150°C; NO partial pressure, 0.03 atm; 25
scans per spectrum*

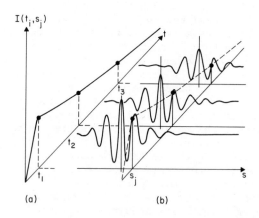

*Figure 15. Relationship between transient source emission at different times and
reconstructed interferograms (37)*

common value of time, t_n, the data are obtained for describing the complete interferogram at t_n. Fourier transformation of the interferogram then provides the spectrum at t_n.

The approach to time resolution of very fast processes described above has been used to study a variety of gas phase chemical reactions stimulated by pulses of either photons or electrons (37). While similar studies have not yet been performed for processes occurring at solid surfaces, it should be feasible to do so. Thus, for example, the dynamics of adsorption and reaction at a single crystal surface, could be studied using a chopped molecular beam as the source of reactants. Alternatively, a pulsed photon beam could be used to repeatedly stimulate a reaction between adsorbed species.

Conclusions

The present review has shown that under equivalent operating conditions, FT spectrometers can offer significantly superior performance over grating spectrometers. The optical advantage of an FT spectrometer can be realized as either a higher signal to noise ratio for an equivalent data acquisition time, or a lower data acquisition time for an equivalent signal to noise ratio. A major portion of this advantage arises from the fact that an FT spectrometer observes all spectral elements simultaneously and as a result, the overall advantage is highest when the number of resolution elements observed is large. Additional advantage is derived from the high optical throughput of FT spectrometers, particularly at high wavenumbers. However, this advantage is partially compensated by the higher noise equivalent power associated with the TGS detector.

As illustrated by the examples discussed here, the use of FT spectrometers for the observation of surface structures is favored by situations in which the flux of radiation coming from the sample is very low or the data acquisition time is limited. Such cases arise in transmission spectroscopy using strongly absorbing or scattering samples, specular and diffuse reflectance spectroscopy from opaque samples, and emission spectroscopy from low temperature sources. FT spectroscopy is also well suited for observing the dynamics of surface species during adsorption, desorption, and reaction.

Acknowledgment

This work was supported by the Division of Chemical Sciences, Office of Basic Energy Sciences, U. S. Department of Energy, under contract No. W-7405-ENG-48.

Literature Cited

1. Little, L. H., "Infrared Spectra of Adsorbed Species";
    Academic Press: New York, 1966.
2. Hair, M. L., "Infrared Spectroscopy in Surface Chemistry";
    Marcel Dekker: New York, 1967.
3. Kiselev, A. V.; Lygin, V. I., "Infrakrasnye Spektry
    Poverkhnostnykh Soedinenii i Adsorbirovannykh Veshchestv";
    Nauka: Moscow, 1972.
4. Bell, R. J., "Introductory Fourier Transform Spectroscopy";
    Academic Press: New York, 1972.
5. Geick, R., in "Topics in Current Chemistry", No. 58;
    Springer Verlag: Berlin, 1975; p. 73.
6. Griffiths, P. R., "Chemical Infrared Fourier Transform
    Spectroscopy": Wiley; New York, 1975.
7. Ferraro, J. R.; Basile, L. J., Eds., "Fourier Transform
    Infrared Spectroscopy - Applications at Chemical Systems",
    Vol. 1; Academic Press: New York, 1978.
8. Ferraro, J. R.; Basile, L. J., Eds., "Fourier Transform
    Infrared Spectroscopy - Applications to Chemical Systems",
    Vol. 2; Academic Press: New York, 1979.
9. Griffiths, P. R.; Sloane, H. J.; Hannah, R. W., Appl. Spec.,
    1977, 31, 485.
10. Mattson, D. R., Appl. Spec., 1978, 32, 335.
11. Tai, M. H.; Harwitt, M., Appl. Opt., 1976, 15, 2664.
12. Low, M. J. D.; Goodsel, A. J.; Takezawa, N., Env. Sci.
    Technol., 1971, 5, 1191.
13. Goodsel, A. J.; Low, M. J. D.; Takezawa, N., Env. Sci.
    Technol., 1972, 6, 268.
14. Low, M. J. D.; Jacobs, H.; Takezawa, N., Water, Air, Soil
    Pollut., 1973, 2, 61.
15. Low, M. J. D.; Lee, P. L., Water, Air, Soil, Pollut., 1973,
    2, 75.
16. Low, M. J. D.; Yang, R. T., J. Catal., 1974, 34, 479.
17. Low, M. J. D.; Goodsel, A. J.; Mark, H., in "Molecular
    Spectroscopy 1971", Hepple, P., Ed.; Inst. of Petroleum;
    London, 1972.
18. Bouwman, R.; Freriks, I. L. C., Appl. Surf. Sci., 1980, 4,
    11.
19. Bouwman, R.; Freriks, I. L. C., Appl. Surf. Sci., 1980, 4,
    21.
20. Low, M. J. D.; McManus, J. C., Chem. Comms., 1967, 1166.
21. Harkness, J. B. L., Ph.D. thesis, M.I.T., 1970.
22. Apse, J. I., Ph.D. thesis, M.I.T., 1973.
23. Bailey, R. B.; and Richards, P. L., "Infrared Absorption
    Spectroscopy of Surfaces: A Low Temperature Detection
    Scheme"; Lawrence Berkeley Laboratory Report LBL-7639;
    Berkeley, CA, 1979.
24. Willey, R. R., Appl. Spec., 1976, 30, 593.
25. Fuller, M. P.; Griffiths, P. R., Anal. Chem., 1978, 50, 1906.

26. Klier, K., "Investigation of Adsorption Centers, Molecules, Surface Complexes, and Interactions Among Catalyst Components by Diffuse Reflectance Spectroscopy", this volume.
27. Bates, J. B., in "Fourier Transform Infrared Spectroscopy - Applications to Chemical Systems", Vol. 1; Ferraro, J. B.; Basile, L. J., Eds.; Academic Press: New York, 1978.
28. Low, M. J. D.; Abrams, L.; Coleman, I., Chem. Comms., 1965, 389.
29. Low, M. J. D.; Coleman, I., Spectrochim. Acta, 1966, 22, 369.
30. Low, M. J. D., J. Catal., 1965, 4, 719.
31. Kember, D.; Sheppard, N., Appl. Spec., 1975, 29, 496.
32. Kobayashi, H.; Kobayashi, M., Catal. Rev., 1974, 10, 139.
33. Tamara, K.; Onishi, T., Appl. Spec. Rev., 1975, 9, 133.
34. Bennett, C. O., Catal. Rev., 1976, 13, 121.
35. Ueno, A.; Bennett, C. O., J. Catal., 1978, 54, 31.
36. Savatsky, B. J., and Bell, A. T., unpublished results.
37. Durana, J. F.; Mantz, A. W., in "Fourier Transform Infrared Spectroscopy - Applications to Chemical Systems", Vol. 2; Ferraro, J. B.; Basile, L. J., Eds.; Academic Press: New York, 1979.

RECEIVED June 3, 1980.

Organic Monolayer Studies Using Fourier Transform Infrared Reflection Spectroscopy

D. L. ALLARA

Bell Laboratories, Murray Hill, NJ 07974

Vibrational spectroscopy of adsorbed surface species is a rapidly developing field. Although the traditional method has been infrared spectroscopy a number of competing techniques now exist including surface enhanced Raman, infrared surface wave, electron energy loss, inelastic electron tunneling and neutron scattering. These methods are treated in detail elsewhere[1] and will not be discussed further here. Transmission infrared spectroscopy[1] is one of the simplest methods to use experimentally but is insufficiently sensitive for detecting monolayers on smooth surfaces. The reflection infrared techniques on the other hand are sufficiently sensitive for flat monolayers. Internal reflection spectroscopy (IRS) is quite sensitive because of the large number of multiple reflections which can be used. However, the method requires a suitable transparent infrared material for the reflection element which also serves as the basic support for the sample. In some cases this provides obvious difficulties in studying phenomena such as adsorption on bulk metals. The subject of internal reflection has been reviewed in detail by Harrick.[2] External reflection spectroscopy (ERS) is useful for detecting spectra on reflective substrates,[3] such as bulk metals, thus is an obvious complement to internal reflection. The most popular use of ERS to date has been the study of CO on metals.[1] In contrast there are very few reports of studies of monolayers of larger organic molecules on metals and metal oxides. Francis and Ellison were among the pioneers in the field with their report of spectra from oriented monolayers of barium stearate on metal mirrors.[4] Since that time there have been a number of reported spectra of thin films of organic molecules on metal mirrors but often the films have been many monolayers thick with no measurement of the actual thickness consequently these reports have not really involved examination of isolated surface species. Some studies which appear to involve very thin films, if not monolayers, have dealt with formic acid on copper[5],[6],[7] aluminum[6] and nickel[7] and nitric oxide and isoamyl nitrite on copper, nickel and iron.[6] The first study above[5] indicates that formic acid only chemisorbs in the presence of oxygen but other results with 10^{-9} torr vacuum suggest otherwise.[7] Earlier work on the chemisorption of acetic acid on copper[8] (with a cuprous oxide overlayer) indicates that adsorption of carboxylic acids can build up vacuum stable, multilayer films so any oxide formed on the copper in the above studies[5],[6] could result in multilayer coverages. Boerio and Chen[9] have obtained spectra from ~15Å thick films of an epoxy polymer deposited onto iron and copper surfaces (oxide covered). These authors[10] have also reported spectra of monolayer thickness films of long chain fatty acids and alcohols on iron and copper surfaces (oxide covered). A study of ethylene adsorption on silver and platinum surfaces in

0-8412-0585-X/80/47-137-037$05.00/0
© 1980 American Chemical Society

which spectra were examined as a function of coverage has been reported.[11]

In contrast to the minimal activity in infrared reflection studies the technique of inelastic electron tunneling spectroscopy (IETS) recently has contributed a large amount of information on monolayer adsorption of organic molecules on smooth metal oxide surfaces,[1],[12] mostly thin (20-30Å) aluminum oxide layers on evaporated aluminum. These results indicate that a variety of organic molecules with acidic hydrogens, such as carboxylic acids and phenols chemisorb on aluminum Oxide overlayers by proton dissociation[13],[14],[15] and that monolayer coverage can be attained quite reproducibly by solution doping techniques.[16] The IETS technique is sensitive to both infrared and Raman modes.[12] However, almost no examples exist in which Raman[1] and or infrared spectra have been taken for an adsorbate/substrate system for which IETS spectra have been observed.

The present study was initiated to provide a direct comparison of IETS and IR spectra for an identical molecule adsorbed on an aluminum oxide covered, evaporated aluminum substrate. Further, it was of interest to see if a weakly acidic C-H bond, such as that present in 1,3-dialkanediones, would show dissociative chemisorption similar to the well-known chemisorptions of Bronsted acids containing acidic O-H bonds (see above). The molecules chosen for this study were acetic acid and 2,4-pentanedione. Both oxide covered copper and aluminum were used as substrates in order to see the effects of substrate oxide on the chemisorption spectra.

Theory

A brief description of the physical principles of the reflection experiment is in order because of the significant differences between optimum experimental conditions and theoretical calculations for reflection and the more usual technique of transmission. A representation of the reflection experiment is given in Figure 1. The theory was originally developed in detail by Hansen[17] and Greenler.[18] The principles developed for the IR indicate[18] that maximum signal strength of monolayers on a metallic substrate should be obtained for a p-polarized beam of light incident at near glancing angles. These conditions give the highest value of the surface E field. The latter almost completely consists of the Z component and consequently only oscillators which have a Z component to their transition dipole moment will absorb energy at their excitation frequencies. This selection rule can be of use in determining the orientation of adsorbed molecules although there have been no quantitative demonstrations of this reported. For optimum conditions a reflection spectrum of a thin film on a metal surface is generally 10-50 times stronger than the corresponding transmission spectrum obtained (at normal incidence) for the same film supported on a transparent substrate. Quantitative calculations of reflection phenomena can be made from the boundary value solutions of Maxwell's equations for the interaction of a propogating infinite plane wave with a system of parallel layers.[17],[18] A discussion of the theoretical and experimental aspects of bond intensities and shifts for thin films on reflective surfaces has been reported[19] recently. In general, strong bands are likely to show some distortion in reflection spectra. For monolayers this usually amounts to a small upward shift in frequency of the band maximum.[19] For example, a band with a line width (at half height) of 25 cm^{-1} should show a shift of $\sim +9$ cm^{-1}. This type of shift must be taken into account when carefully comparing IR reflection spectra of monolayers with high resolution vibrational spectra obtained by other means.

Experimental

Sample Preparation

Metal films were prepared by thermal deposition of pure metals (>99.99%) at pressures in the 10^{-8} torr range using polished Si substrates (surface roughness ~50Å). Film thicknesses were measured with a quartz thickness monitor and were generally about 200 nm. After deposition the vacuum system was backfilled with pure oxygen gas. The samples were quickly removed, flooded with ethanolic solutions of the selected organic adsorbate (usually 0.01 M) and spun horizontally at several thousand RPM using a standard photoresist spinner. These are representative of the standard conditions used for preparation of IETS samples by solution doping.[12] Once formed the films were quite stable to further exposure to the ambient environment.

Spectra

Spectra were obtained using a Digilab 15-B Fourier transform infrared spectrometer operating at 2 cm^{-1} resolution. A diagram of the optical system is shown in Fig. 2. The source output is processed through a mirror and aperture system (A) to give a parallel beam of light entering the interferometer. The exiting beam is stopped down through a variable aperture (1-2 cm) (A') and processed through another mirror system to give an ~f60 beam focussing to a 2 mm beam spot (determined by A) at a position where the center of the sample is placed. the s-component is removed by a polarizer placed at P and/or P'. The reflected beam is finally focused onto a liquid nitrogen cooled mercury cadmium telluride detector. The general details of Fourier transform infrared spectroscopy are discussed elsewhere.[1] All spectra were taken in a thoroughly dry, N_2 purged atmosphere and samples were stored between spectra in tightly closed, individual fluoro-polymer containers. The angle of incidence (ϕ, see Fig. 1) was 86 degrees for all the reported spectra.

Results and Discussion

Acetic Acid

Acetic acid chemisorption has been previously studied using IETS by Lewis, Mosesman and Weinberg[14] for oxide covered aluminum surfaces. Using reflection IR Tompkins and Allara[8] have reported spectra for adsorption on oxidized copper and Hebard, Arthur and Allara[20] for adsorption on oxidized indium. All these studies demonstrate that chemisorption from the gas phase involves proton dissociation since the observed spectra are those of acetate ion species.

The present IR results with solution doping are shown in Figs. 3 and 4 for adsorption onto oxide covered aluminum and copper, respectively. For aluminum oxide the peaks at 1590, 1405 and 1335 cm^{-1} match up in frequency with the major IETS peaks reported[14] at 1589, 1403 and 1331 cm^{-1}. However, the intensity pattern is significantly different since the weakest of the above peaks in IETS is at 1589 cm^{-1} whereas this is the strongest IR peak. The IETS peaks, respectively, have been attributed to[14] C=0 and C-0 stretching vibrations, C-H asymmetric deformation and C-H symmetric deformation. Although partly a matter of terminology, we attribute the IETS peak at 1589 cm^{-1} to the asymmetric carboxylate stretch. The IETS peak at 1452 cm^{-1} should then correspond to the symmetric stretch. The appearance of the 1452 cm^{-1} mode only in IETS is possible on the basis that its symmetry makes it only weakly

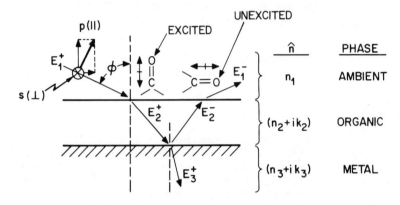

Figure 1.   Description of the single reflection experiment. The $C = O$ oscillator is shown to demonstrate the surface selection rule.

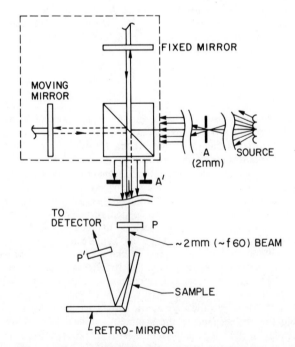

Figure 2.   Simplified diagram of the optical layout of the FTIR spectrometer; A and A′ are apertures, P and P′ are polarizers, and the area inside the dashed box is the actual interferometer assembly

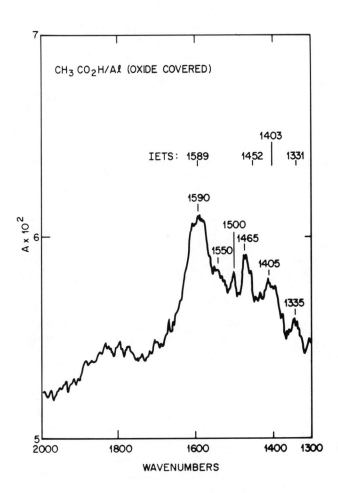

Figure 3. Reflection spectrum of an oxidized aluminum film exposed to 0.010M acetic acid in ethanol; also shown are the IETS peak positions (13)

IR active but Raman active and IETS selection rules allow Raman modes.[12] However, since these stretching modes probably do not have ideal symmetry near the surface one might expect additional IR intensity for the symmetric mode. Another factor to be considered is the orientation. If the acetate ions were oriented with their rotational symmetry (C_{2v}) axis normal to the surface, the surface selection rule would favor the symmetric mode with its dipole derivative perpendicular to the surface in contrast to the asymmetric mode with its dipole derivative parallel to the surface and thus the symmetric/asymmetric ratio would be enhanced. Most IETS evidence[12] suggests that carboxylate ions are oriented away from the aluminum oxide surface with variable angles of tilt, as the structure in Fig. 5 suggests and accordingly some symmetric/asymmetric enhancement would be expected. It seems unlikely that the 1465 cm^{-1} peak in in the present experiments is the 1452 cm^{-1} symmetric carboxylate IETS mode shifted up in frequency although some small optical up shifts could occur in the reflection experiment[19] but certainly less than 13 cm^{-1}. It is possible that the 1465 cm^{-1} peak in the IR spectrum could correspond to a CH_2 bending mode of a co-adsorbed ethoxide species formed by adsorption of ethanol which is used as the solvent for the acetic acid solution. Evans and Weinbeg[21] report an IETS peak at 1472 for adsorbed ethanol on aluminum oxide at ambient temperatures and attribute it to the CH_2 bending mode. However, the 1465 cm^{-1} peak does not occur for adsorption on copper or for adsorption of 2,4-pentanedione on aluminum oxide, although a weak 1460 cm^{-1} peak is present in the latter, which suggests that co-adsorption of ethanol may not be the correct explanation for the above acetic acid results. The current evidence suggests that a different acetate species (or mixture of species) is observed in the present experiments than was observed by IETS. This conclusion is strengthed by the appearance of the 1500 cm^{-1} peak which is unobserved in IETS and which is presently unassigned to any specific mode. Further the 1550 cm^{-1} shoulder observed in Fig. 3 may be the asymmetric stretching mode of a second surface species of acetate ion (see below).

The IR spectrum for adsorption on oxidized copper (Fig. 4) exhibits a different relative intensity pattern than the spectrum for the oxidized aluminum case although the peak positions are roughly the same. These peaks match up in frequency with the major peaks reported for the gas-phase adsorption on thick Cu_2O films[8] except for the 1630 cm^{-1} peak [22] which is not observed in the present study. The gas-phase results showed that vacuum stable multilayer films were formed and that the species present varied from run to run as evidenced by the variability in the intensity of the 1630, 1590 and 1550 cm^{-1} peaks. The spectrum shown in Fig. 4 is probably that of a monolayer or near-monolayer coverage judged by the weak intensity, $A = -log(R/R_o) < 10^{-2}$. Thus the process of coverage by solution deposition and rapid spinning does not seem to allow multilayer films to form permanently. The 1555-1591 cm^{-1} peaks can be assigned to the asymmetric carboxylate stretch of different acetate species and the 1405 cm^{-1} shoulder to the asymmetric C-H deformation. The 1448 cm^{-1} band would most likely correspond to the symmetric carboxylate stretch (see discussion above for the aluminum oxide case) except that it is uncharacteristically strong relative to the asymmetric mode. However, this may be due to the orientational effects discussed above. The present results thus would suggest that the orientation on oxidized copper films may be closer to normal to the surface than on oxidized aluminum. Exposure of the films to the acetic acid solution for hours instead of seconds gives roughly the same results, as seen in Fig. 4, except the peaks are not as well resolved. Obviously thick multilayer films, if formed in solution, do not remain after removing the excess solution.

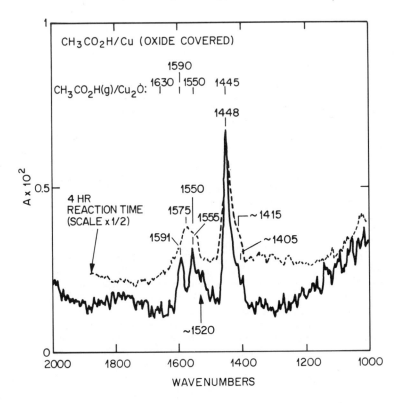

Figure 4.   Reflection spectrum of an oxidized copper film exposed to 0.010M acetic acid in ethanol. The dashed line is the spectrum after a 4-h exposure of a fresh substrate to the acetic acid solution. Also shown are the IR peak positions from the gas-phase exposure spectrum (7).

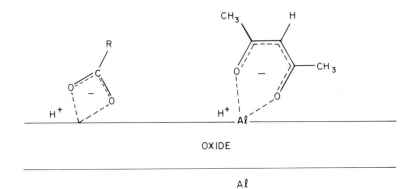

Figure 5.   Representation of possible structures for adsorbed carboxylate and acetylacetonate ions on aluminum oxide

To these two sets of results we can contrast the IR spectrum reported for adsorption from several torr of acetic acid gas onto oxidized indium films.[20] This spectrum consists of two broad peaks at $\sim$1590 and 1455 cm^{-1}, attributed to the asymmetric and symmetric carboxylate stretch.[20] Both peaks are of roughly similar intensity.

It is obvious from the above results that adsorption of acetic acid, and, of course, presumably other carboxylic acids, is different in detail from one metal oxide to another and is perhaps also somewhat a function of whether adsorption occurs from gas or solution phase. However, in all cases acetate ions are formed and differences presumably reflect more subtle features of surface structure and chemistry. In general, there seems to be a correspondence between the frequencies reported by IR and IETS for IR active modes although intensity patterns are not similar, as one should expect based on the different mechanisms of vibrational excitation. Further work is obviously needed to define the differences between the two spectroscopies more exactly.

2,4-Pentandione

Adsorption of 2,4-pentanedione (acetylacetone, AcAcH) on oxidized aluminum gives rise to the reflection IR spectrum shown in Fig. 6. The corresponding transmission spectra for the pure diketone and aluminum acetylacetonate ($Al(AcAc)_3$) are also shown. The reflection spectrum shows peaks at 1610, 1535, 1460 (weak) and 1405 cm^{-1} and also a strong, broad peak at $\sim$950 cm^{-1} due to the aluminum oxide phonon.[22] The high frequency modes (1730, 1710 cm^{-1}) present in the AcAcH transmission spectrum are absent in reflection and generally there is little direct correspondence between the spectra of the pure and adsorbed material. On the other hand there is a better match between the peak positions of the adsorbed species and the $Al(AcAc)_3$ complex. This strongly suggests that AcAcH forms the $AcAc^-$ enolate ion on chemisorption and that some sort of aluminum acetylacetonate complex is formed on the surface as depicted in Fig. 5, the exact orientation and bonding of such a complex is presently undetermined and Fig. 5 is only meant to convey a possible type of structure.

Chemisorption on oxidized copper gives a roughly similar spectrum to that on oxidized aluminum but with a down shifted high frequency peak, a new peak at 1435 cm^{-1} and no observable peak at 1460 cm^{-1}, as shown in Fig. 7. The transmission spectrum of the copper acetylacetonate complex exhibits major peaks at 1577, 1552, 1529, 1461, 1413, 1353 and 1274 cm^{-1} in the spectral range of Fig. 7. Again the absence of the "free" C=0 high frequency stretch modes of the pure diketone indicate that an $AcAc^-$ enolate type of structure is adsorbed but the failure of the surface spectrum to match the $Cu(AcAc)_3$ spectrum indicates that the surface species is not simply a layer of the complex salt. Exposure of a fresh substrate to the adsorption solution for 4 hours gives no changes in the reflection spectrum except for some loss of resolution and a suggestion of a shoulder forming at $\sim$1580 cm^{-1}. Thus film growth does not appear to occur under these conditions. The intensity of the peaks in Fig. 7 (-log $(R/R_o) < 10^{-2}$) indicate that approximately monolayer coverages are obtained by the solution adsorption techniques.

Relationship to Catalysis

Clearly, well resolved vibrational spectra of monolayer coverages of organic species can be obtained by reflection IR as well as IETS (although IETS exhibits a greater sensitivity). The ability to study chemisorption on smooth metal surfaces with

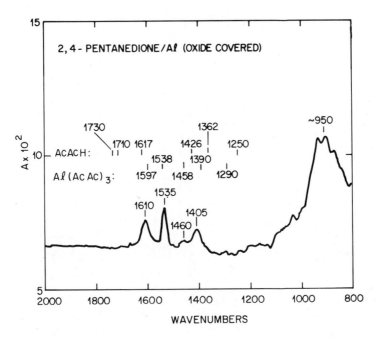

Figure 6.   Reflection spectrum of an oxidized aluminum film exposed to 0.010M 2,4-pentanedione in ethanol. Also shown are the peak positions for transmission spectra of the pure diketone (AcAcH) and the aluminum complex (Al(AcAc)₃).

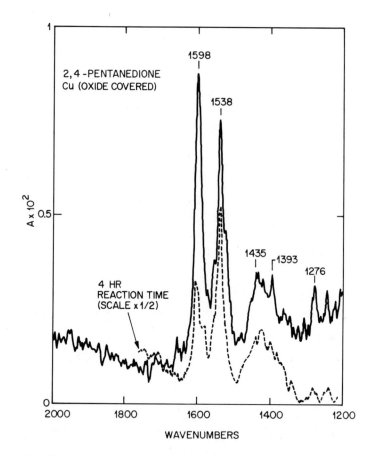

Figure 7. *Reflection spectrum of an oxidized copper film exposed to 0.010M 2,4-pentanedione in ethanol. The dashed line is the spectrum after a 4-h exposure of a fresh substrate to the ketone solution.*

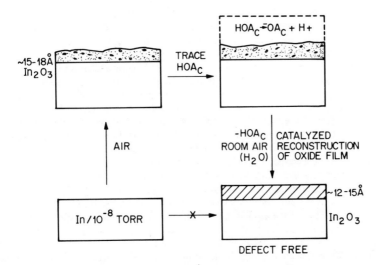

Figure 8. Overall description of the acetic acid catalyzed reconstruction of a defect containing indium oxide overlayer on indium (20)

ambient overlayers, usually oxides, under various conditions of temperature and pressures make reflection IR an obvious tool for examining catalytic mechanism of organic surface reactions. As mentioned earlier in this paper several pioneering studies have been reported. Both the results in this paper and those reported in IETS studies (see earlier sections) for aluminum oxide are of interest with respect to a number of recent reports by Posner *et al* on the catalysis of a variety of liquid-phase organic reactions on aluminum oxide surfaces. [23] Such reactions include intramolecular additions of OH and acidic CH groups, intermolecular additions of X-H groups where X is 0, S, Se, N, C, etc. and various oxidation-reduction and substitution reactions as well as other reactions. A number of these appear to involve weakening of acidic X-H bonds by the alumina surface to formally give reactions of the X^- species. The spectral studies mentioned in the present paper are quite relevant since they establish firm evidence for the likelihood of such important ionized intermediates in appropriate cases of catalysis over alumina. For example, acidic C-H groups of enolizable ketones can lead to substitution reactions at the acidic C atom position. The present results show that 1,5-pentandione exists in some sort of enolic structure on an aluminum oxide surface. Another example is catalysis of epoxide ring opening by acetic acid to give acetate substituted products. The presence of surface adsorbed acetate ions is most likely under these experimental conditions where alumina surfaces are exposed acetic acid solutions.

Another interesting and different type of catalysis is involved in the catalyzed reconstruction of an indium oxide overlayer on indium. [20] This study was alluded to earlier in the discussion of acetate ion species formed on indium oxide by chemisorption from several torr of acetic acid gas. At low partial pressures of acetic acid ($<< 0.1$ torr) the reversible adsorption of acetic acid catalyzes the reconstruction of a thin ($~ 10\text{-}15\text{Å}$), porous indium oxide overlayer to a defect-free (no pin holes) film as judged by pinhole sensitive tunnel junction measurements. Some clues as to the mechanism were obtained from IR plus Auger and electron loss spectroscopy as well as ellipsometry measurements. The overall process is shown in Fig. 8. This is an example where processes in the substrate themselves can be usefully catalyzed.

The above discussion is meant to point out specific possible application of surface vibrational spectroscopy to new areas of catalysis. Certainly there are many others and brevity prevents further discussion of such a large subject. Reflection IR, IETS and perhaps Raman, which is rapidly developing in useful directions, would appear to have a good future as high resolution techniques for studies of the chemisorption of organic molecules on a variety of substrates.

Conclusions

External reflection infrared spectroscopy appears to have a useful role in future surface studies of adsorbed organic molecules as judged from the increasing number of papers published in the literature. It is capable of examining the same samples as inelastic electron tunneling, although with lower sensitivity, but it has the advantage of flexibility with regard to choice of substrate. The present paper shows application to both aluminum oxide/aluminum and cuprous oxide/copper substrates for solution adsorption of acetic acid and 2,4-pentanedione. Both molecules adsorb by proton dissociation of the most acidic proton to form ionic surface species. The species from acetic acid adsorption differs slightly from that detected by inelastic tunneling for the case of gas phase adsorption. The characterization of fairly complicated organic molecules at surfaces by vibrational techniques offers a powerful means of examining the nature of a variety of surface reactions ranging from general organic synthesis in metal oxides to specific phenomena such as catalyzed oxide overlayer reconstruction.

References

[1] See other papers in this symposium.

[2] N. J. Harrick in "Characterization of Metal and Polymer Surfaces", Ed. by L. H. Lee, Academic Press, NY, 1977, vol. 2, p. 153.

[3] For a general review of the technique see: H. G. Tompkins in "Methods of Surface Analysis", Ed. by A. W. Czanderna, Elsevier, NY, 1975, Chapter 10.

[4] S. A. Francis and A. H. Ellison, J. Opt. Soc. Amer. *49* 131 (1959).

[5] R. W. Stobie and M. J. Dignam, Can J. Chem *56* 1088 (1978); these authors used ellipsometric techniques rather than reflectance to obtain spectra.

[6] M. Ito and W. Suetaka, J. Phys. Chem *79* 1190 (1975).

[7] M. Ito and W. Suetaka, J. Catalysis *54* 13 (1978).

[8] H. G. Tompkins and D. L. Allara, J. Coll Interface Science *47* 697 (1974).

[9] F. J. Boerio and S. L. Chen., Applied Spectroscopy *33* 121 (1979).

[10] F. J. Boerio and S. L. Chen, J. Coll. Interface Science, in press.

[11] M. Ito and W. Suetaka, Surf. Science *62* 308 (1977).

[12] See for example, P. K. Hansma, Physics Reports, Section C of Physics Letters, *30* 145 (1977).

[13] J. Klein, A. Leger, M. Belin, D. Deforneau and M. J. L. Sangster, Phys. Rev. B7 2336 (1973).

[14] B. F. Lewis, M. Mosesman and W. H. Weinberg, Surface Science, *41* 142 (1974).

[15] B. F. Lewis, W. M. Bowser, J. L. Horn, Jr., T. Luu and W. H. Weinberg, J. Vac. Sci. Technol. *11* 262 (1974).

[16] J. D. Lanagan and P. K. Hansma, Surface Science *22* 211 (1975).

[17] See W. N. Hansen in "Progress in Nuclear Energy", vol. 11, H. A. Elion and D. C. Steward Eds., Chapter 1; and in "Advances in Electrochemistry of Electrochemical Engineering", vol. 9, P. Delahay and C. W. Tobias, Eds., Wiley, New York, 1973, pp. 1-60 and references therein.

[18] R. G. Greenler, J. Chem. Phys. *44* 310 (1966).

[19] D. L. Allara, A. Baca and C. A. Pryde, Macromolecules *11* 1215 (1978).

[20] A. F. Hebard, J. R. Arthur and D. L. Allara, J. Appl. Physics. *49* 6039 (1978).

[21] H. E. Evans and W. H. Weinberg, J. Chem. Phys. *71* 1537 (1979).

[22] F. P. Mertens, Surf. Science *71* 161 (1978).

[23] For a general review see: G. H. Posner, Angew Chem. Int. Ed. Engl. *17* 487 (1978).

RECEIVED June 16, 1980.

Intermolecular Interactions and the Infrared Reflection–Absorption Spectra of Chemisorbed Carbon Monoxide on Copper

P. HOLLINS and J. PRITCHARD

Chemistry Department, Queen Mary College, Mile End Road, London E1 4NS, United Kingdom

The structure of the exposed metal surface in a supported catalyst cannot be readily established by direct methods, but the possibility of obtaining information about the metallic surface from the infrared spectra of adsorbed molecules was evident in the early work of Eischens, Francis and Pliskin (1). They reported coverage dependent changes in the transmission infrared spectra of CO chemisorbed on silica-supported palladium, nickel and platinum. Effects were particularly prominent in the low frequency bands of bridge-bonded CO on Ni and Pd. The latter showed additional sub-peaks developing with increasing coverage. Eischens et al took this as evidence for a heterogeneous surface composed of a limited number of relatively homogeneous parts which could have been the major crystal faces of Pd. Their interpretation of the spectra was supported by the later work of Baddour, Modell and Goldsmith (2) who found that the distribution of intensity within the broad envelope of the band of bridge-bonded CO on Pd could be modified by using the supported Pd as a catalyst for CO oxidation. After aging until a stable level of catalytic activity was reached ("break in"), the intensity distribution had shifted towards higher frequencies. Such changes may be plausibly attributed to the development of preferred crystallite facets under reaction conditions. This example illustrates the potential that the infrared spectroscopy of suitable adsorbates may offer for the investigation of the structures of supported catalyst surfaces, but in order to extract topographical information the spectrum associated with each specific surface structure must be identified. For the facets on relatively large particles one may hope to obtain reference spectra from adsorbates on macroscopic faces of single crystals by reflection-absorption infrared (RAIR) spectroscopy (3, 4, 5, 6). In this paper we discuss the application of this approach to the investigation of copper surfaces and the extent to which it is justified in the light of recent interpretations of the frequency shifts caused by intermolecular interactions in adsorbed layers.

0-8412-0585-X/80/47-137-051$06.00/0
© 1980 American Chemical Society

Infrared Spectra of CO Adsorbed on Polycrystalline Copper

Transmission infrared spectra of CO adsorbed on supported copper have been reported frequently since the original investigation by Eischens, Pliskin and Francis (7). A single intense band attributed to linearly-bonded CO groups is usually found between 2100 and 2130 cm^{-1}. Figure 1 shows an example of the growth of the band with increasing coverage on alumina-supported copper. As the intensity increases the band broadens but it neither shifts in frequency nor develops any new structure. The small difference in the reported positions of the band when silica or alumina is the support (Table I(a)) may depend to some extent on particle size effects (for very small particles) or on the completeness of reduction of the copper surface. Co-adsorbed oxygen for example can displace the band by 10 to 20 cm^{-1} (8). Much better reproducibility has been attained in RAIR spectra from polycrystalline copper films deposited under ultrahigh vacuum conditions (Table II(a)), the peak frequency lying between 2102 and 2106 cm^{-1}. The problem of reduction does not arise, and the particle sizes are large in copper films deposited at room temperature. Until results from single crystal surfaces became available it seemed unlikely that significant information about the structure of the copper surface would be gained from such simple and reproducible spectra. One might have expected CO adsorbed on any of the major crystal faces to give a band at just over 2100 cm^{-1}.

The first single crystal results were therefore surprising (9,10), a band at 2085 cm^{-1} being reported for Cu(100) and a band at 2076 cm^{-1} for Cu(111). The absence of appreciable absorption at these positions in the spectra of CO on silica- or alumina-supported copper, or in the RAIR spectra of CO on copper films deposited on glass (11,12) and on tantalum ribbons (13), led to the unexpected conclusion (4) that the major low index faces were conspicuous by their scarcity in polycrystalline copper surfaces.

Subsequent studies (8) of CO adsorption on copper single crystals included the (110), (311) and (755) surface orientations, and a reinvestigation of the (100) surface led to a revised band position of 2079 cm^{-1} at low coverage which changed to 2088 cm^{-1} at high coverage (14). The bands from the single crystal surfaces are compared in Figure 2 with the band from a polycrystalline film on glass. It should be borne in mind that single crystal spectra have generally been recorded at low temperatures whereas the majority of the spectra of CO on supported copper or on evaporated films have been recorded at room temperature. Cooling alone does not shift the bands, but at low temperatures an additional stage of adsorption occurs which is accompanied by a reduced heat of adsorption, a reduction of the positive surface potential generated in the first stage, and sometimes a small shift of the infrared absorption band. For the comparison in Figure 2, therefore, the positions and shapes of the bands at the completion of the first stage only have been used.

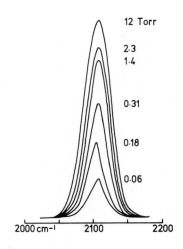

Figure 1. IR band of CO chemisorbed on alumina-supported Cu at room temperature and at the pressures indicated (8)

Table I

Frequencies of the Infrared Band in some Transmission
Spectra of CO Adsorbed on Supported Copper

	Support	Temperature/K	Wavenumber/cm^{-1}	Reference
(a)	Silica	298	2120	(7)
	"	"	2100	(53,54)
	"	113	2110	(55)
	"	298	2103,2105	(8)
	Alumina	298	2100	(54)
	"	298	2110	(55)
	"	298	2108	(8)
(b)	Magnesia	298	2081	(8)

Table II

Frequencies of the Infrared Band in RAIR Spectra
from Evaporated Copper Films

	Substrate	Temperature/K	Wavenumber/cm^{-1}	Reference
(a)	Glass	298	2105	(11)
	"	298	2102	(8)
	"	298	2102	(40)
	"	77	2105	(12)
	Tantalum foil	298	2105	(13)
	Alumina	298	2106	(8)
(b)	Magnesia	298	2082	(8)

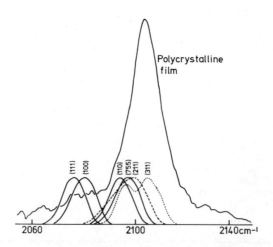

*Figure 2.   Comparison of IR band of CO chemisorbed on a polycrystalline Cu film
at room temperature with the bands at the completion of the first stage of adsorption
at low temperatures on single crystal Cu surfaces of the indicated orientations (6)*

·· It is clear that the CO spectra from the more open (110) face
and the stepped (211), (311) and (755) faces are much closer to
the spectra generally obtained from polycrystalline films or
supported copper. A significant exception to this correlation was
found when MgO was used to support dispersed copper or as the substrate for polycrystalline films (Tables I(b) and II(b)). The
peak of the CO absorption band then occurred at $\sim$ 2080 cm^{-1} which
coincides with the bands from the low index faces. The cube face
of MgO, which is an ionic crystal with the NaCl structure, is well
known as a substrate on which epitaxial metal films will grow with
low index orientations. Thus it seems likely that the pronounced
difference between copper on magnesia on the one hand and copper
on glass or amorphous oxide supports on the other is associated
with the presence or absence of preferred orientations in the
nucleation stages of growth (8). On MgO the low index orientations of the nuclei persist through the stages of coalescence and
aggregation, apparently even through films of 400 nm thickness
(8), but on the amorphous supports randomly oriented nuclei have
no cause to reorientate during coalescence. Measurements of the
anisotropy of the surface energy of copper (15) show it to be at
most 2% at 1200 K. The development of a smooth surface of minimum
area could then be more important than reorientation for deposits
on amorphous substrates, leading to surfaces with a wide variety
of orientations including continuously variable step spacings for
curved surfaces. Thus for structure sensitive catalytic reactions
the control of the nucleation and growth of the metal particles
may be an important function of the support quite apart from any
chemical role it may play.

Surface Coverage Effects in the Infrared Spectra of Chemisorbed CO

Most studies of the infrared spectra of CO adsorbed on
supported copper have been limited to room temperature and to
pressures at which the surface coverage would be less than the
saturation coverage. The actual coverage would differ between one
facet and another, depending on the relative heats of adsorption.
Consequently the preceding comparison of the spectra of CO on
supported copper catalysts and evaporated films with the spectra
from single crystal surfaces has been greatly facilitated by the
remarkable insensitivity of the band positions to surface coverage. This is illustrated by the spectra of CO on Cu(100) shown in
Figure 3 (14).
The very small changes in the position and width of the band
are all the more remarkable because of the structural changes revealed by low energy electron diffraction. During the second
stage of adsorption the overlayer structure changes from c(2 x 2)
to a compression structure in which CO molecules were thought to
be out of registry with the copper surface net (16,17). Despite
this, the band of linearly bonded CO hardly alters. However, the
LEED pattern of the compression structure is susceptible to an

alternative interpretation (18). The alternative overlayer
structure is composed of bands of linearly bonded CO molecules
occupying sites on top of copper atoms, separated by rows of
molecules in two-fold bridging sites. In this model adjacent
bands are in antiphase and each band has a c(2 x 2) structure.
It was suggested that the latter property could account for the
small influence of compression on the infrared spectrum in the
linear region. This interpretation would have been strongly
supported by any spectroscopic evidence for the rows of bridge-
bonded CO molecules, but careful examination by high resolution
electron energy loss spectroscopy (19,20) has failed to show any
additional feature in the CO stretching frequency range.

A similar problem arises with the spectra of CO on Cu(111)
shown in Figure 4. Once again LEED evidence shows a sequence of
structures. The completion of the first stage of adsorption at
the surface potential maximum gives a $(\sqrt{3} \times \sqrt{3})$-30⁰ overlayer
structure in which all molecules could occupy equivalent on-top
sites. As the coverage increases to saturation the overlayer
appears to change to a close-packed hexagonal array of CO molecules out of registry with the hexagonal surface net of the
copper substrate. Some broadening of the infrared band accompanies
this change, but the band shifts a mere six wavenumbers from
2076 cm^{-1} to 2070 cm^{-1} (10,21). An alternative interpretation of
the LEED evidence was suggested (18), in which both linear and
bridged CO groups are involved, but the subsequent infrared study
(21) revealed no sign of a second band.

The lack of evidence for anything but linearly bonded CO
groups on Cu(100) and Cu(111) is remarkable in view of the small-
ness of the frequency shifts during the formation of compression
structures. However, for the purpose of comparing spectra from
the various forms of copper at room temperature we should be more
concerned with the first stage of CO adsorption. Here too the
frequencies are almost constant, but in comparison with the
spectra of CO adsorbed on other metals such constancy is the exception rather than the rule. It is not found on platinum, for
example, where the band of linearly adsorbed CO is similar both
in position and in intensity to that on copper, and where the
surface potential varies with coverage much as it does on copper
(22,23). Instead, the band shifts from 2063 cm^{-1} at low coverage to 2100 cm^{-1} at high coverage on Pt(111) (23,24,25), an effect
similar to the shift from 2040 to 2070 cm^{-1} found with supported
platinum (1). Large upward frequency shifts have also been found
for the lower frequency bands of bridged CO on nickel (26) and
palladium (27) single crystal surfaces. On the other hand, the
high frequency bands of linearly bonded CO on silver (12) and
gold (12,28) show downward shifts.

Such shifts can be rationalized in terms of the model of
synergic σ and π bonding of the CO ligand in metal carbonyls (29)
as was proposed by Blyholder (30). The formation of a strong
surface bond involves considerable back donation from metal d

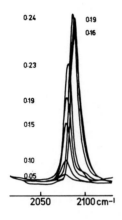

Figure 3.   Spectra of CO on Cu(100) at coverages indicated by the surface potential values (V); SP increasing (left) in Stage 1 and decreasing (right) in Stage 2 (14)

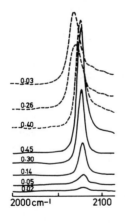

Figure 4.   Spectra of CO on Cu(111) at increasing coverages, indicated by the surface potential values (V) left; (———) SP increasing in the first stage towards the maximum value at $\Theta = 1/3$, (– – –) then decreasing in the second stage towards saturation at $\Theta = 0.52$ (21)

orbitals into the antibonding $2\pi^*$ orbital of the CO molecule, leading to a weakening of the C-O bond and a lowering of the C-O stretching frequency. Competition between adsorbed molecules may then result in a reduction of this effect and an increase in the stretching frequency with increasing coverage.

The downward shift that is found with silver and gold starts from a very high frequency. It too can be understood (12) if it is assumed that the weak surface bond on these metals is almost entirely σ in character, involving the slightly antibonding 5σ orbital of the free CO molecule. Dative overlap of the filled 5σ orbital with acceptor orbitals of the metal surface causes a transfer of charge density towards the metal, enhancing the C-O bond and its stretching frequency as well as contributing to the positive surface potential. With increasing coverage the accumulation of surface charge may become self-inhibiting, weakening the surface bond and lowering the C-O frequency. The heat of adsorption of CO on Ag(111), initially small, falls almost linearly with increasing coverage (31). Alternatively, charge accumulation could be relieved by a small but increasing degree of back donation into the $2\pi^*$ orbital as the coverage increases. Back donation is expected to be very restricted in comparison with the transition metals because of the depth of the d bands below the Fermi level in gold and particularly silver.

Effects of this kind, involving modification of the surface bond by competition of adsorbate molecules, will be termed "chemical" as distinct from physical effects such as dipole coupling. Some evidence for their existence is provided by studies of co-adsorbed electron donor and electron acceptor ligands (32,33). Within the framework of this qualitative chemical interpretation of coverage induced frequency shifts it is clearly possible to imagine a case where the balance between the opposing charge transfers in σ and π bonds is such that no competition develops as the coverage increases. CO on copper might appear to be such a case. However, even if the bonding is unaffected by coverage a shift towards higher frequencies is generally expected because of local field or dipole coupling effects.

The upward frequency shift of the band of CO on supported platinum was originally connected with dipole dipole interactions by Eischens, Pliskin and Francis (1) because of the coverage dependent behaviour of the spectra of a mixture of ^{12}CO and ^{13}CO. At low coverages the mixture gave two bands with an isotopic separation of some 50 cm^{-1} as in the gas phase. With increasing coverage, however, the lower frequency ^{13}CO band remained essentially fixed in position while the ^{12}CO band shifted to higher frequencies. Furthermore, the ^{12}CO band gained disproportionate intensity at high coverages at the expense of the ^{13}CO band. Hammaker, Francis and Eischens (34) subsequently developed a mathematical treatment of dipole dipole interactions between chemically identical chemisorbed molecules that could account for these experimental observations. More recently the same theory has been employed by Crossley and King to account for frequency shifts in the RAIR spectra of CO on Pt{111} (24,35).

Dipole Coupling

Treating the adlayer as an array of N parallel oscillating point dipoles Hammaker et al (34) expressed the vibrational potential energy V as

$$2V = \sum_{i=1}^{N} \lambda_i^1 Q_i^2 + 2 \sum_{i>j=1}^{N} R_{ij}^{-3}(\partial\mu/\partial Q_i)(\partial\mu/\partial Q_j)Q_iQ_j \tag{1}$$

where Q_i, Q_j are the normal coordinates of the vibrations of molecules i and j, distant R_{ij} apart, and $(\partial\mu/\partial Q)$ represents the derivative of the dipole moment μ with respect to the normal coordinate. The first term corresponds to the uncoupled molecules, and λ^1 is related to the wavenumber ω_1 of the vibration of an isolated adsorbed molecule (singleton) by $\lambda^1 = 4\pi^2c^2\omega_1^2$. The second term describes the dipole coupling in which the dynamic dipole moments $(\partial\mu/\partial Q)Q$ interact with a R^{-3} dependence. Coupling effects, leading to a perturbation of the original frequency, arise whenever the potential energy expression includes a term of the form $\sum a_i a_j Q_i Q_j$ involving the products of the normal coordinates of two molecules. Dipole coupling is one such case; vibrational coupling via the surface bonds and the metal has also been considered recently (36) and can lead to similar consequences.

For an isotopically pure adsorbate in which one molecule i is surrounded by (N-1) identical molecules j eqn(1) leads to the following exact solution for the only infrared active mode (34),

$$\lambda = \lambda_i^1 + (\partial\mu/\partial Q)^2 \sum_{j=2}^{N} R_{ij}^{-3} \tag{2}$$

where the perturbed frequency ω is given by $\lambda = 4\pi^2c^2\omega^2$. For later purposes it will be convenient to replace $(\partial\mu/\partial Q)$ by the vibrational polarizability α_v to which it is related by

$$\alpha_v = (\partial\mu/\partial Q)^2/4\pi^2c^2\omega_1^2 \tag{3}$$

Thus eqn(2) may be expressed alternatively as

$$(\omega/\omega_1)^2 = 1 + \alpha_v T \tag{4}$$

where T is the dipole sum $\sum_j R_{ij}^{-3}$. Since T increases with coverage the perturbed frequency also increases.

Provided the overlayer structure is known T can be calculated and α_v may be evaluated from the observed frequency shift. On Pt(111) the adsorption of CO gives a c(4 x 2) structure (22) in which half the CO molecules are considered to be linearly bonded (37). Crossley and King (35) calculated T = 0.059 $\overset{o}{A}^{-3}$ for the linearly adsorbed molecules in the structure and their experimental frequency shift from 2063 cm^{-1} to 2100 cm^{-1} gives $\alpha_v = 0.61$ Å^3. [The value of $d\mu/dr = 9.8 \times 10^{-20}$ C given by Crossley and King for the derivative of μ with respect to displacement r appears to be an underestimate due to the omission of a factor 4π from $4\pi\epsilon_0$ in their eqn(1). The corrected value is 3.5×10^{-19} C, corresponding to an effective charge $e^* = 2.2e$]. That the whole shift could be attributed to dipole coupling and contains no appreciable chemical contribution was considered by Crossley and King to be shown by the spectra of mixtures of ^{12}CO with ^{13}CO.

Hammaker et al (34) deduced approximate expressions for the frequencies of the two infrared active modes of the system composed of a central adsorbed molecule of one isotopic species coupled to an environment of the other species. In one mode the labelled molecule vibrates in phase with its neighbours giving a frequency ω_h higher than the frequency ω_ℓ of the other mode where the motion is 180° out of phase. The two frequencies are related to ω_1' (the frequency of the 2-D lattice in the absence of the labelled molecule) and ω_2 (the frequency of the labelled molecule in the absence of surrounding molecules, i.e. a labelled singleton) by

$$(\omega_h/\omega_1')^2 = 1 + \alpha_v^2 \sum_{ij} R_{ij}^{-6} \omega_2^2 (\omega_1'^2 - \omega_2^2) \tag{5}$$

$$(\omega_\ell/\omega_2)^2 = 1 - \alpha_v^2 \sum_{ij} R_{ij}^{-6} \omega_1'^2 (\omega_1'^2 - \omega_2^2) \tag{6}$$

The terms in α_v^2 in these equations are very much less than the term in α_v of eqn(4). Physically this is a consequence of the coupling between oscillators of different frequency being very much weaker than between those of the same frequency. Hammaker et al estimated the coupling effect to be an order of magnitude less. Thus in a dilute isotopic mixture the minor component is effectively decoupled from its surroundings.

Crossley and King (35) pointed out that this provides a method of distinguishing between chemical and dipole coupling contributions to the total shift seen with a single isotopic species. For the c(4 x 2) overlayer structure of CO on Pt(111), ω_h was estimated to be only 2 cm^{-1} greater than the singleton frequency of ^{12}CO. Thus if the shift to 2100 cm^{-1} in the complete layer of ^{12}CO is due entirely to dipole coupling, the dilution of the ^{12}CO with ^{13}CO, while keeping the chemical composition the same, should reduce the frequency back to 2065 cm^{-1}. This result was obtained, convincingly demonstrating the insignificance of any coverage dependent chemical shift. It should also be emphasised that coupling produces a strong transfer of intensity from the low frequency band to the high frequency band in the spectra of the mixture, an effect originally observed by Eischens et al (1) and also shown very clearly in the results of Crossley and King.

The success of the isotope dilution experiment for CO on
Pt(111) was accompanied by a serious difficulty in reconciling the
magnitude of the shift, which determines $(\partial\mu/\partial Q)$, with the intensity of the band, which also determines $(\partial\mu/\partial Q)$. When due
allowance is made for the resultant surface field and geometric
factors (36) in RAIR spectroscopy the intensity is almost consistent with the vibrational polarizability $\alpha_v = 0.057$ Å^3 (39),
corresponding to the gas phase intensity, as has been concluded
for CO adsorbed on copper films (40) from infrared studies and for
CO on Pt(111) (41) and Cu(100) (42) from high resolution electron
energy loss spectroscopy. This value of α_v is an order of magnitude smaller than that deduced from the frequency shift.
Moskovits and Hulse (36) therefore proposed that the larger part
of the frequency shift of CO on platinum may be due to a "through-
metal" coupling mechanism operating in addition to the "through-
space" dipole coupling. Such vibrational coupling has been mentioned briefly above, and as it affects the vibrational potential
energy in a way similar to dipole coupling it can generate
frequency shifts and also intensity transfers. However, the
coupling constant is no longer predictable as it is for dipole
coupling. Consequently large effects could occur for a given
integrated band intensity. Moskovits and Hulse suggested that the
much smaller shifts found for CO on copper surfaces may well
reflect a weaker vibrational coupling because of the weaker
surface bond. Unfortunately, independent evidence for vibrational
coupling in adsorbed layers has not been reported, so the effect
remains speculative. But once the role of the metal surface is
included it becomes possible to explain large shifts even by
dipole coupling.

Image Effects in Dipole Coupling

The theory of Hammaker et al treated the adlayer in isolation
and also included only the vibrational part of the molecular
polarizability. Important modifications were made by Delanaye,
Lucas and Mahan (43) and by Mahan and Lucas (39) who (i) introduced the classical image plane at a distance d below the point
dipole layer (Figure 5), (ii) included the electronic polarizability α_e of the adsorbed molecules, and (iii) corrected for non-
classical dielectric screening. The result of (i) is that the
polarization of the metal by the fields of the dipoles is as if
the polarized metal were replaced by image dipoles, equal and
opposite to the real dipoles, at a distance d below the image
plane. To this approximation, therefore, the influence of the
metal on dipole coupling is to add an image dipole sum V to the
real dipole sum T introduced previously. Even an isolated induced dipole μ interacts with its self-image to give a potential
energy $-\mu^2/4d^3$. Scheffler (44) pointed out that the correct way to
combine the effects of the self-image and the images of the other
dipoles is to include the self-image term with the sum V. Since

the self-image plays a major part in modifying the dipole coupling theory it is convenient to make its effect explicit by defining V to include all images except the self-image and then to add in the self-image term separately. For the single isotope case Scheffler deduced in place of eqn(4) the result

$$(\omega/\omega_1)^2 = 1 + \alpha_v(S - 1/4d^3)/[1 + \alpha_e(S - 1/4d^3)] \tag{7}$$

in which the dipole sum S is given by

$$S = T + V = \Sigma R_{ij}^{-3} - \sum_{images} 2P_2(\cos\phi)(R_{ij}^2 + 4d^2)^{-3/2} \tag{8}$$

and the screening effect due to the electronic polarizability α_e is allowed for. Since the image dipoles subtend different angles ϕ at the reference dipole it is necessary to include the Legendre polynomials $P_2(\cos\phi)$ in (8). Eqn(7) differs from Mahan and Lucas' result in the presence of the terms $-1/4d^3$.

The distance d, which is the separation between the point dipole and the image plane, presents some difficulty because it is by no means obvious where the image plane lies. If one takes a simple chemical bonding approach in which

$$d = \frac{1}{2}(C-O \text{ bond length}) + \text{typical M-C bond length} - \text{metallic radius}$$

a value close to 1 Å is obtained. More sophisticated treatments recognise that the image concept is a crude approximation at atomic distances and that d is a parameter to be chosen to give the best possible fit to the actual field (44,45,46). In principle, a value of d can be obtained from LEED intensity measurements (47), but all calculations of dipole shifts carried out to date have treated d simply as an adjustable parameter.

An important difference between eqn(7) and eqn(4) is that ω_1 no longer represents the frequency of an adsorbed singleton but rather the frequency of a hypothetical adsorbed molecule isolated from its self-image too. The singleton frequency is given by the limit of eqn(7) as $S \rightarrow 0$:

$$\omega(\text{zero coverage}) = \omega_1[1 - \alpha_v/(4d^3 - \alpha_e)]^{\frac{1}{2}} \tag{9}$$

As d is reduced ω would reach zero when

$$4d^3 = \alpha_e + \alpha_v = \alpha_s$$

corresponding to the distance at which the image field provides positive polarization feedback leading to the polarization catastrophe. Physically reasonable values of d for adsorbed molecules lie very close to this critical distance, and ω(zero coverage) is then a rapidly varying function of d (see Figure 6). Since $\alpha_e \gg \alpha_v$ the critical condition (10) is approximately

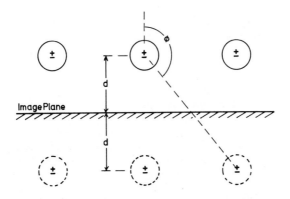

Figure 5.   Diagram of dipoles and image dipoles

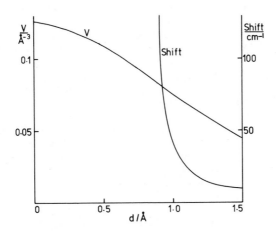

Figure 6.   Dependence on distance d *from image plane of the image dipole sum* V *for the* $(\sqrt{3} \times \sqrt{3})$*–30° structure (*$\Theta = 1/3$*) of CO on Cu(III). Also shown is the frequency shift,* $\Theta = 0 \rightarrow \Theta = 1/3$*, which depends strongly on the displacement of the singleton frequency to lower values by the self-image.*

equivalent to $\alpha_e/4d^3 = 1$. Thus, for reasonable values of d, eqn(7) shows that a small change in S may produce a large change in ω, and large coverage dependent frequency shifts can be accounted for. Figure 6 shows the smaller effect of d on the image dipole sum V in a typical overlayer structure.

Scheffler also pointed out that α_s is an anisotropic tensor and that the appropriate value to be chosen for chemisorbed CO is the component parallel to the molecular axis (2.60 $\mathring{A}^3$) rather than the spatially averaged value (1.95 $\mathring{A}^3$) used previously. Using this revised value, together with suitable choices for ω_1 and d, he was able to fit both Crossley and King's (35) data for CO on Pt(111) and Bradshaw and Hoffmann's results for CO on Pd(100 (27).

Before discussing the interpretation of recent results for CO on Cu(111) (21) we must first consider the case of mixed isotopes in the light of the modifications which have been made to the theory of Hammaker et al (34). Equations (5) and (6) are still relevant to the limiting situation of a molecule of one isotopic species completely surrounded by molecules of the other, but two modifications must be made, both of which are analogous to the changes already made for the single isotope case; the sum over R_{ij}^{-6} must be extended to include image dipoles other than the self-image, and the effect of α_e on the effective vibrational polarizability must be included. When these alterations are made the formulae yield shifts only of the order of 0.1 cm^{-1}, compared with typical single isotope shifts of 20 cm^{-1} or more. The original model of Hammaker et al predicted a shift of about 2 cm^{-1}; this comparatively large value was a consequence of the high value of α_v which had to be assumed in order to give the correct value for the single isotope shift.

Spectra of $^{12}CO-^{13}CO$ Mixtures on Cu(111)

We recall that the coverage dependent frequency shifts on copper surfaces are small. In the light of the preceding discussion possible explanations for this difference from the behaviour of CO on platinum and palladium include:

(i) Coupling is very weak. The actual magnitude of the dipole coupling shifts on platinum and palladium have not been established with certainty. A major part of the observed shift may conceivably be due to vibrational coupling and vibrational coupling on copper may be much weaker, as suggested by Moskovits and Hulse (36).

(ii) The monolayer may develop by an island growth mechanism such that a singleton frequency is never observed. This behaviour has been proposed (23) to account for the growth of the band of CO on Pt(111) at 80 K at a constant frequency of 2090 cm^{-1} instead of shifting from the singleton frequency of 2063 cm^{-1} as at higher temperatures. However, no evidence for the early development of ordered islands of CO on copper has been found in low temperature LEED studies,

nor have larger frequency shifts been found in the infrared
spectra at higher temperatures.

(iii) A large dipole coupling shift may occur on copper, similar
to the shifts observed on platinum, but it may be balanced
by a chemical shift in the opposite direction. Such a
balance of opposing shifts was proposed long ago by Van
Hardeveld and Van Montfoort (48) to account for the almost
constant frequency of N_2 on nickel, and suggested again by
Crossley and King (35). The variability from one copper
surface to another of the size and sign of the small shifts
that do occur would be consistent with imperfect balancing.

These alternatives can be distinguished readily by the spectra
of $^{12}CO-^{13}CO$ mixtures which are dilute in one component.

(i) If coupling is weak the spectra should show two bands with
intensities proportional to the relative abundances of the
isotopes at all coverages and with a constant frequency
difference of about 47 cm^{-1} as in the gas phase.

(ii) If coupling is strong but islands are formed even at low
average coverage, the low coverage bands should show
anomalous intensity ratios and separations different from
47 cm^{-1}.

(iii) If a chemical shift to lower frequency balances a coupling
shift to higher frequency, the band of the minor isotopic
component, which is effectively decoupled from surrounding
CO molecules, should show the chemical shift to lower
frequency, while the major band should behave like that of
the pure isotopic adsorbate.

Spectra have been obtained as a function of CO coverage on
Cu(111) at 82 K for several isotopic compositions (21). Two sets
of coverage dependent spectra are shown in Figure 7, one for
mixtures containing only 29% of ^{13}CO, the other for mixtures containing only 3% of ^{12}CO. They demonstrate quite unambiguously that
the third explanation is correct. Island formation is disproved
by the 47 cm^{-1} frequency separation and the relative intensities
at low coverage, both of which change as the coverage increases.
The transfer of intensity to the high frequency band is particularly marked at high coverage in the 3% ^{12}CO spectra. Coupling
is clearly strong, but the minority bands show a large downward
chemical shift. For the major component the chemical shift is
offset by the coupling shift. Similar behaviour is expected on
other faces of copper.

Coupling shifts in the ordinary CO spectra can now be determined as a function of coverage. The range of coverage up to the
maximum surface potential is particularly interesting. It
corresponds to the completion of the first stage of CO adsorption
and to the formation of the first ordered structure, $(\sqrt{3} \times \sqrt{3})-30°$,
at a coverage $\theta = 1/3$. Only one kind of binding site is believed
to be occupied in this coverage range. Furthermore, there is good
evidence (49) that the sticking probability of CO at 82 K is constant up to $\theta = 1/3$, so that the coupling shifts can be correlated

with coverages estimated from relative exposures.

Scheffler's model was first tested by calculating the value of d which would account for the coupling shift of 25 cm^{-1} at $\theta = 1/3$, using the polarizabilities corresponding to gas-like molecules as discussed above. With the value $\alpha_v = 0.057$ Å^3, the shift could be fitted by d = 1.07 Å and S = 0.198 Å^{-3}. The shift is very sensitive to the choice of d, much more so than the image dipole sum V, because of the strong influence of the self-image on the singleton frequency (eqn(9)). The shift and V are shown in Figure 6 for the $(\sqrt{3} \times \sqrt{3})$-30°overlayer on Cu(111).

The shifts at lower coverage then depend on the model adopted for the packing of molecules in the adlayer. Partial coverage may correspond to a dilated version of the full coverage structure, in which case S is approximately proportional to $\theta^{3/2}$. Alternatively, adsorption sites may be randomly occupied until the final structure is completed, and Mahan and Lucas (39) have shown that S will then be approximately proportional to θ. In this case there will also be a slight broadening of the band, but it will usually be indetectable compared with the natural band width. The shifts predicted by these models do not differ greatly, and it is very likely that intermediate behaviour might occur in which nearest neighbour sites would tend not to be occupied but in which no long-range order exists. In Figure 8 the experimental results are compared with both the S $\propto \theta$ and S $\propto \theta^{3/2}$ models. While neither extreme is clearly favoured it is evident that the dipole coupling theory can account for the experimental findings.

The surface potential also arises from surface dipoles and it is interesting to compare its coverage dependence with the coverage dependence of the infrared band frequency. The mutual depolarization of parallel adsorbed dipoles with increasing coverage was treated by Topping (50). More recently Bradshaw and Scheffler (51) have included the effect of image dipoles, concluding that the overall result of the dipole interactions can be described by a factor $[1 + \alpha_e(S - 1/4d^3)]^{-1}$ (or Q) such as occurs in the frequency shift expression, eqn(7). The surface potential should then be proportional to Q x θ, and Figure 9 shows that this is approximately true for CO on Cu(111) up to $\theta = 1/3$ if the same value of d = 1.07 Å is assumed. However, it is by no means obvious that d should be the same for the centres of the charge distributions involved in the static surface dipole (SP) and the dynamic surface dipole (IR). Indeed, consideration of the infrared band intensities throws doubt on the appropriateness of both d = 1.07 Å and $\alpha_v = 0.057$ Å^3 for the dynamic dipole.

Figure 9 shows that the infrared band intensities vary with coverage in a similar way to the surface potential, i.e. apparently as Q x θ. [We note in passing that the infrared bands shown in Figure 4 are accompanied by a baseline offset between the low and high frequency sides. This is an artefact which has been allowed for. It is a result of normalising the derivative spectra before integration by ratioing with the reflected in-

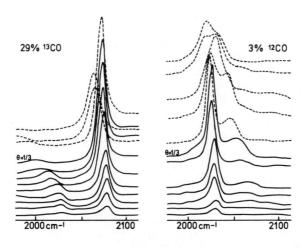

Figure 7.   Spectra of ^{12}CO–^{13}CO mixtures on Cu(111) at 82 K coverages increasing (left) from bottom to top; (——) first stage; (– – –) second stage (21)

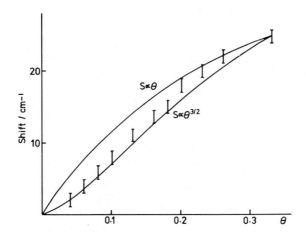

Figure 8.   Dipole coupling shifts (experimental points) as function of coverage in Stage 1, compared with theoretical shifts for random (S $\propto$ Θ) and regular (S $\propto$ $\Theta^{3/2}$) distributions assuming free molecule polarizabilities

tensity spectrum of the clean surface rather than with the covered surface. It is apparent because the exceptional intensity of the bands introduces a large difference between the intensity spectra across the derivative features in the wavelength modulated spectra]. Such a coverage dependence of the intensity is not of the expected form. According to Scheffler (44) the effect of the dipole interactions can be expressed by an effective molecular polarizability $\alpha^*(\omega,\theta)$ related to the polarizability $\alpha(\omega)$ of the free molecule by $\alpha^* = \alpha[1 + \alpha(S - 1/4d^3)]^{-1}$, and the contribution of each molecule to the intensity of the reflection-absorption band is proportional to the imaginary part of α^*. The latter proves to be proportional to Q^2, but it is clear from Figure 9 that the experimental intensities deviate considerably from the predicted $Q^2 \times \theta$ relationship.

If both the coverage dependence of the frequency shift and of the intensity are to be reconciled throughout the coverage range $0 < \theta < 1/3$ the values of both d and α_v must be increased. The integrated band intensities of terminal carbonyl groups in the metal carbonyls (38) are generally much larger than the integrated intensity of gaseous CO. Consequently a larger value of α_v appears to be very reasonable for chemisorbed CO and is certainly indicated by estimates from transmission spectra of CO adsorbed on supported metals (38). Figure 10 shows that a satisfactory agreement of theory and experiment can be achieved with $\alpha_v = 0.18 \text{ Å}^3$ and d = 1.85 Å, leaving α_e unchanged. The larger value of d is in much better accord with the position of the surface barrier on copper estimated from LEED intensity analyses (47).

It is assumed here that the parameters α_e, α_v and d can be regarded as independent of coverage. In view of the simultaneous chemical shift with coverage one cannot rule out the possibility the coupling shift could be affected by slight changes in these parameters. If the image distance were to decrease slightly with coverage as a result of the redistribution of surface charge density, it would increase the effect of the self image field which produces a downward frequency shift, estimated now (with d = 1.85 Å) as 8 cm^{-1} for a singleton. The self image effect is opposed at higher coverages by the upward shifts caused by interactions with other dipoles and their images. Although both of these opposing shifts depend on d only the upward shift is affected by isotopic decoupling. Thus the chemical shift may contain a contribution from a changing value of d.

In summary, the point dipole model with images can be made to account for the experimentally determined effects of intermolecular dipole coupling. The magnitudes of the effects cannot be predicted from the properties of the free CO molecule but they can be used to estimate the changed values in the chemisorbed state.

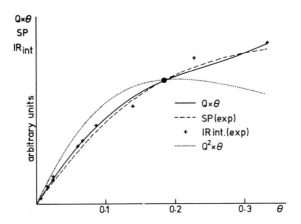

*Figure 9.   Coverage dependence of the surface potential and of the IR band inten-
sity of CO on Cu(111) compared with Q × Θ and Q × Θ² where Q is based on the
polarizabilities of the free CO molecule and d = 1.07Å. All curves have been normalized to pass through the circled point.*

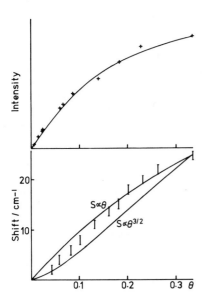

*Figure 10.   (Top) experimental values
(×) of the relative IR band intensity compared with Q² × Θ, where Q is based on
$\alpha_v = 0.18$Å³ and d = 1.85Å; (bottom)
coverage dependence of the dipole coupling shift (experimental points) compared
with theoretical shifts for random (S ∝ Θ)
and regular (S ∝ Θ³/²) distributions, assuming the same values of α_v and d, and
with $\omega_I = 2086$ cm⁻¹*

Implications of Dipole Coupling for the Interpretation of
Transmission Infrared Spectra of Adsorbates on Supported Metals

It is now clear that the very small frequency shifts observed in the spectra of CO on copper surfaces are a result of
the fortuitous cancelling of two much larger but opposing shifts.
Dipole coupling probably occurs to a significant extent with CO
on silver and gold surfaces too, despite the lower coverages
which can be achieved (21). The appreciable overall downward
shifts on these metals suggest an even larger chemical shift than
on copper, in keeping with ideas about the relative involvement
of σ and π bonding (12).

Dipole coupling is a long range effect, whereas the chemical
effect may be limited to a few lattice spacings. The balance between the combination of short range and long range interactions
could then be seriously disturbed if the dipole sums are
truncated by adsorption being limited to a small area. The
latter could occur on the facets of small metal crystallites in a
supported catalyst, and if the result is large enough it may in-
validate the use of RAIR spectra from macroscopic single crystal
faces as reference spectra for the interpretation of transmission
spectra from catalysts.

It is important, therefore, to consider the rate at which
the dipole coupling shift at a reference molecule increases towards its limiting value as nearest neighbours, next nearest
neighbours, etc., are successively taken into account. The model
system used here consists of a small hexagonal island of the
$(\sqrt{3} \times \sqrt{3})$-30° overlayer of CO on Cu(111).

For the three smallest islands - containing 7, 19 and 37 CO
molecules - the full vibrational eigen value problem was solved
to yield dipole shifts relative to the singleton of 12, 17 and
20 cm^{-1} respectively. In addition, it was found that for these
small islands certain vibrational modes other than the fully in-
phase mode have non-zero intensity, and in each case bands
having an intensity of a few percent of the major band's appear
at lower frequency.

For larger islands no attempt was made to solve the problem
exactly, but instead the shift corresponding to the dipole sum
at the central molecule was evaluated. These shifts are shown as
a function of the island's maximum diameter in Figure 11. The
calculation slightly over-estimates the size of the shift which
would be observed for these hexagonal islands, but should give a
reasonable indication of the shift to be expected for an
irregular island whose "typical" radius is ½ D. It is seen that
although the dipole sum is very slow to converge fully (not
having reached its final value at the centre of a cluster of a
thousand CO molecules), the initial convergence is very rapid
indeed, and only very small islands will show marked differences
from the infinite layer. If one assumes that only variations
greater than 5 cm^{-1} (half the width of the band) are significant,

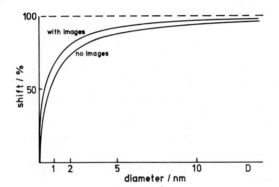

Figure 11.   The dependence on island size D *of the dipole coupling shift relative to that in an infinite layer of (√ 3 × √ 3)–30° structure*

then no significant difference will be observed for particles
greater than approximately 3 nm in diameter. This conclusion is
not much affected by the neglect of image interactions.

A 5 cm^{-1} shift is rather less than the difference between the
IR frequencies observed on different faces of copper at comparable
coverages, so incomplete dipole sums are not a major factor in
determining the CO vibrational frequency on moderately large
particles. It is interesting to note, however, that a difference
of approximately 5 cm^{-1} has been observed between the CO frequencies on alumina and silica-supported Cu (8) and that the higher
frequency does correlate with the larger particle size. The much
larger difference between the spectra from low index faces and the
stepped surfaces of copper gives good justification for the
original interpretation of the transmission spectra from films and
dispersions supported on amorphous and crystalline substrates (8).

The discussion of interactions has concentrated on the first
state of adsorption. The nearest neighbour distance in the
$(\sqrt{3} \times \sqrt{3})$-30° structure is already 4.4 Å, not very much greater
than the van der Waals diameter of a CO molecule (3.3 Å), and one
may well doubt the validity of the point dipole model for
quantitative calculations of interactions at such coverages, quite
apart from the uncertainties surrounding the image approximation.
It is clear from Figure 7 that both the coupling shift and the
chemical shift, whatever their actual forms may be, continue to
increase as the coverage increases towards saturation. Under
these conditions the local charge distribution must surely be
taken into account, and following Dows (52) the van der Waals
interaction between the individual atoms of adjacent vibrating
molecules should be included. There is even evidence for lateral
electronic interaction between CO molecules at high coverages on
Pd(100) (51).

Acknowledgement

We gratefully acknowledge the support of this work by the
Science Research Council and the Central Research Fund
Committee of London University.

Literature Cited

1.   Eischens, R. P., Francis, S. A. and Pliskin, W. A., _J. Phys._
      Chem., 1956, 60, 194.
2.   Baddour, R. F., Modell, M. and Goldsmith, R. L., _J. Phys._
      Chem., 1970, 74, 1787.
3.   Tompkins, H. G., in Czanderna, A. W., Ed. "Methods of
      Surface Analysis", Elsevier, New York, 1975.
4.   Pritchard, J., in "Moderne Verfahren der Oberflächenanalyse",
      Dechema-Monogr., 1975, 78, 231.
5.   Pritchard, J. and Catterick, T., in Anderson R. B. and
      Dawson, P. T., Ed. "Experimental Methods in Catalytic

Research", Vol. 3, Academic Press, New York, 1976.

6. Pritchard, J., in Roberts, M. W. and Thomas J. M., Ed. "Chemical Physics of Solids and their Surfaces", Vol. 7, Chemical Society Specialist Periodical Reports, London, 1978.

7. Eischens, R. P., Pliskin, W. A. and Francis, S. A., J. Chem. Phys., 1954, 22, 1786.

8. Pritchard, J., Catterick, T. and Gupta, R. K., Surface Sci., 1975, 53, 1.

9. Chesters, M. A., Pritchard, J. and Sims, M. L., Chem. Comm., 1970, 1454.

10. Pritchard, J., J. Vac. Sci. Technol., 1972, 9, 895.

11. Bradshaw, A. M., Pritchard, J. and Sims, M. L., Chem. Comm., 1968, 1519.

12. Chesters, M. A., Pritchard, J. and Sims, M. L. in Ricca, F., Ed. "Adsorption Desorption Phenomena", Academic Press, London, 1972, p.277.

13. Tompkins, H. G. and Greenler, R. G., Surface Sci., 1971, 28, 194.

14. Horn, K. and Pritchard, J., Surface Sci., 1976, 55, 701.

15. McLean, M. and Gale, B., Phil. Mag., 1969, 20, 1033.

16. Tracy, J. C., J. Chem. Phys., 1972, 56, 2748.

17. Chesters, M. A. and Pritchard, J., Surface Sci., 1971, 28, 460.

18. Pritchard, J., Surface Sci., 1979, 79, 231.

19. Andersson, S., Surface Sci., 1979, 89, 477.

20. Sexton, B. A., Chem. Phys. Letters, 1979, 63, 451.

21. Hollins, P. and Pritchard, J., Surface Sci., 1979, 89, 486.

22. Ertl, G., Neumann, M. and Streit, K. M., Surface Sci., 1977, 64, 393.

23. Horn, K. and Pritchard, J., J. Phys. (Paris), 1977, 38, C4, 164.

24. Shigeishi, R. A. and King, D. A., Surface Sci., 1976, 58, 379.

25. Krebs, H. J. and Lüth, H., Appl. Phys., 1977, 14, 337.

26. Campuzano, J. C. and Greenler, R. G., Surface Sci., 1979, 83, 301.

27. Bradshaw, A. M. and Hoffmann, F., Surface Sci., 1978, 72, 573.

28. Kottke, M. L., Greenler, R. G. and Tompkins, H. G., Surface Sci., 1972, 32, 231.

29. Cotton, F. A. and Wilkinson, G., Advanced Inorganic Chemistry, 3rd ed., Wiley, New York, 1972, ch.22.

30. Blyholder, G., J. Phys. Chem., 1964, 68, 2772.

31. McElhiney, G., Papp, H., Pritchard, J., Surface Sci., 1976, 54, 617.

32. Queau, R. and Poilblanc, R., J. Catalysis, 1972, 27, 200.

33. Primet, M., Basset, J. M., Mathieu, M. V. and Prettre, M., J. Catalysis, 1973, 29, 213.

34. Hammaker, R. M., Francis, S. A. and Eischens, R. P., Spectrochim. Acta, 1965, 21, 1295.

35. Crossley, A. and King, D. A., Surface Sci., 1977, 68, 528.

36. Moskovits, M. and Hulse, J. E., Surface Sci., 1978, 78, 397.
37. Froitzheim, H., Hopster, H., Ibach, H. and Lehwald, S.,
    Appl. Phys., 1977, 13, 147.
38. Brown, T. L. and Darensbourg, D. J., Inorg. Chem., 1967,
    6, 971.
39. Mahan, G. D. and Lucas, A. A., J. Chem. Phys., 1978, 68, 1344.
40. Stobie, R. W., Rao, B. and Dignam, M. J., Surface Sci., 1976,
    56, 334.
41. Ibach, H., Surface Sci., 1977, 66, 56.
42. Andersson, S., Surface Sci., 1979, 89, 477.
43. Delanaye, F., Lucas, A. A. and Mahan, G. D., Proc. 3rd
    Intern. Conf. on Solid Surfaces, Vienna, 1977, p.477.
44. Scheffler, M., Surface Sci., 1979, 81, 562.
45. Appelbaum, J. A. and Hamann, D. R., Phys. Rev. B, 1972, 6,
    1122.
46. Zaremba, E. and Kohn, W., Phys. Rev. B, 1976, 13, 2270.
47. Price, G. L., Jennings, P. J., Best, P. E. and Cornish,
    J. C. L., Surface Sci., 1979, 89, 151.
48. Hardeveld, R. van and Montfoort, A. van, Surface Sci., 1969,
    17, 90.
49. Hollins, P., Ph.D. Thesis, London University, 1980;
    Hollins, P. and Pritchard, J., to be published.
50. Topping, A. J., Proc. Roy. Soc. London A, 1927, 114, 67.
51. Bradshaw, A. M. and Scheffler, M., J. Vac. Sci. Technol.,
    1979, 16, 447.
52. Dows, D. A., J. Chem. Phys., 1960, 32, 1342.
53. Eischens, R. P. and Pliskin, W. A., Adv. Catalysis, 1958, 10,
    1.
54. Smith, A. W. and Quets, J. M., J. Catalysis, 1965, 4, 163.
55. Kavtaradze, N. N. and Sokolova, N. P., Russ. J. Phys. Chem.,
    1970, 44, 603.

RECEIVED June 3, 1980.

Investigation of Carbon Monoxide on Nickel(100) by Infrared Ellipsometric Spectroscopy

J. D. FEDYK and M. J. DIGNAM

Department of Chemistry, University of Toronto, Toronto, Ontario, M5S 1A1, Canada

The importance of vibrational spectroscopy as a tool for studying molecular processes on solid surfaces is well established (1). As this paper forms part of a symposium on surface vibrational spectroscopy, with emphasis on applications to heterogeneous catalysis, the introductory material is devoted primarily to a consideration of the advantages of IR ellipsometric spectroscopy (IRES) for such applications. The main comparison made is between IRES studies of specularly reflecting surfaces, and conventional transmission IR studies of dispersed systems. Before addressing this matter, however, we present a summary of IRES, indicating its strengths and weaknesses in relation to other reflection vibrational spectroscopies, specifically other forms of IR specular reflection spectroscopy, high resolution electron energy loss spectroscopy (EELS), and surface Raman spectroscopy.

Although IRES has been used to follow surface reactions under conditions close to those typical of heterogeneous catalysis (2), the data chosen for presentation at this symposium deal mainly with the use of IRES as a probe of molecular interactions and local molecular geometry at high surface coverages, matters of interest in unravelling the fundamentals of heterogeneous catalysis.

IRES Spectroscopy

Fundamental Optical Equations. As a review of IRES has been published recently covering this topic (3), we present here only the material necessary for an understanding of the subsequent sections.

For any planar interface which is optically homogeneous in planes parallel to the interface, there exists in general a pair of unique polarization eigenstates for each direction of the

0-8412-0585-X/80/47-137-075$05.75/0
© 1980 American Chemical Society

incident light. If such an interface exhibits uniaxial symmetry
(all directions in the plane of the interface optically equivalent), these polarization eigenstates correspond to plane
polarized light, polarized with the electric vector parallel to (p-
polarized) and orthogonal to (s-polarized) the plane of incidence
respectively. They are unique in that they undergo no change in
polarization state upon reflection, though in general,each
suffers a change in amplitude and phase. These changes are most
conveniently represented by complex valued reflection coefficients, R_p and R_s, which operate as proportionality constants
connecting the incident and reflected complex amplitudes for the
p and s electric vectors, E_p and E_s, i.e.

[1] $E_p^{(r)} = R_p E_p^{(i)}$, $E_s^{(r)} = R_s E_s^{(i)}$

where the superscripts $^{(i)}$ and $^{(r)}$ identify the quantities
connected with the incident and reflected light, respectively.
Representing R_p and R_s as follows,

[2] $R_p = \rho_p \exp(i\delta_p)$, $R_s = \rho_s \exp(i\delta_s)$

where $i = \sqrt{-1}$, then the modulus of R_p, ρ_p, represents the ratio
of the reflected to incident electric vector amplitude, while
the argument of R_p, δ_p, represents the phase change introduced
by the reflection. In this representation, the spatial and
temporal periodicity of E_p for a monochromatic plane electromagnetic wave is represented by

[3] $E_p = E_{po} \exp[i(2\pi\nu t - 2\pi\hat{n} z/\lambda_o + \delta_{po}]$

where z is the position coordinate in the direction of
propagation of the light, E_{po} the amplitude of the electric field
at z = 0, δ_{po} its phase at z = 0 = t, ν and λ_o the frequency
and wavelength in vacuum of the light, while $\hat{n}$ is the complex
valued refractive index of the medium. An equivalent expression
applies for E_s, on substituting s for p in eq. [3]. For a non-
absorbing medium, the complex refractive index $\hat{n}$, is equal to the
refractive index, n. For an absorbing medium, however, $\hat{n}$ is
given by

[4] $\hat{n} = n - ik$.

The absorption index, k, introduces an exponential decay in E_p
(and E_s) with increasing z (see eq. [3]). Recalling that the
intensity, I, is related to the electric field amplitude E,
according to $I \propto |E|^2$, it follows from eqs. [3] and [4] that

[5] $I_z = I_o \exp[-(4\pi k/\lambda_o)z] = I_o 10^{-\beta z}$

where β is the chemical extinction coefficient for the medium and is related to k according to

[6] $\beta = 4\pi k/2.303\lambda_o$

Thus in this representation, the imaginary part of the refractive index takes care of the absorption properties of the medium in question.

Returning for a moment to the reflection coefficients, from eqs. [1] to [3] we have

$$\rho_p = E_{po}^{(r)}/E_{po}^{(i)} \text{ and } \delta_p = \delta_{po}^{(r)} - \delta_{po}^{(i)} ,$$

in accord with the statements following eq. [2].

If the light incident on the interface is fully polarized (not necessarily plane polarized, however), then the electric vector may be represented by a sum of vectors in the p- and s-directions,

[7] $\underline{E}^{(i)} = E_p^{(i)}\underline{p} + E_s^{(i)}\underline{s}$,

where $\underline{p}$ and $\underline{s}$ are unit vectors. In general, the vector resultant $\underline{E}$, of adding $E_p\underline{p}$ and $E_s\underline{s}$ will trace out, with time, an elliptical figure at each position, z, i.e. elliptically polarized light results. If E_p and E_s have the same phase angle ($\delta_{po} = \delta_{so}$), then the elliptical figure collapses to a line (i.e., linear polarized light results making an angle $\tan^{-1}(E_{so}/E_{po})$ with the p-axis). For $\delta_{po} - \delta_{so} = \pm\pi/2$, and $E_{po} = E_{so}$, circularly polarized light results.

Upon reflection, $\underline{E}^{(i)}$ is transformed to $\underline{E}^{(r)}$ according to

[8] $\underline{E}^{(r)} = R_p E_p^{(i)}\underline{p} + R_s E_s^{(i)}\underline{s} = R_s[\underline{E}^{(i)} - (1-R_p/R_s)E_p^{(i)}\underline{p}]$.

Thus for $R_p/R_s = 1$, the incident and reflected light, while they may differ in phase and amplitude, have the same polarization states (i.e., $\underline{E}^{(r)}$ and $\underline{E}^{(i)}$ will trace out figures in the s-p plane which differ from one another by a scaling factor only). For a given wavelength and angle of incidence, the ratio R_p/R_s completely characterizes the polarizing properties of the interface at that wavelength. It is conventional to represent the polarizing properties of an interface by angles ψ and Δ defined according to

[9] $R_p/R_s = (\tan\psi)\exp i\Delta = (\rho_p/\rho_s)\exp[i(\delta_p - \delta_s)]$.

For convenience in handling reflectance absorbances, which are defined in terms of logarithms of intensities, we define the following quantities in terms of the logarithms of the reflection coefficients,

[10] $L_p = -\ln R_p$, $L_s = -\ln R_s$, $L = L_p - L_s = -[\ln(\tan\psi) + i\Delta]$.

For p-polarized light incident upon our test surface, the
reflectance absorbance for adsorbed species, A_p, is defined
according to

[11] $A_p = \log_{10}(\bar{I}_p^{(r)}/I_p^{(r)}) = \log_{10}|\bar{R}_p/R_p|^2$

$$= (2/2.303)[(Re(L_p) - Re(\bar{L}_p)] = (2/2.303)Re(\Delta L_p)$$

where the bar over the symbol designates either a clean or
reference state condition for the surface, and Re denotes the
real-part-of. An analogous expression applies for s-polarized
light.

Reflectance ellipsometry is concerned with the measurement
of the polarizing properties of an interface, i.e., with the
measurement of $\tan\psi$ and Δ, or of (R_p/R_s) for the "clean" and
covered surface. Defining the ellipsometric reflectance
absorbance according to

$$A_e = \log_{10}[|\bar{R}_p/\bar{R}_s|^2/|R_p/R_s|^2]$$

it follows that

[12] $A_e = (2/2.303)Re(\Delta L) = A_p - A_s$

where $\Delta L = L - \bar{L}$.

This is the fundamental equation connecting ellipsometric
reflection spectroscopy with photometric reflection spectroscopy. Thus the ellipsometric absorbance spectrum, (A$_e$ versus
λ_o) is simply the reflectance absorbance for p-polarized light,
less that for s-polarized light. For highly reflecting substrates
such as metals in the I.R., the requirement that the tangential
component of the electric field vector be continuous across the
interfacial region leads to E_s being essentially zero at the
interface (since it is near zero in a highly reflecting medium,
exactly zero in a perfect conductor, $|\hat{n}| = \infty$). For metallic
substrates in the I.R., which are the only systems that we shall
be concerned with, $A_e \simeq A_p$ so that photometric and ellipsometric
spectroscopies yield the same absorption spectrum. What then is
the advantage of ellipsometric over photometric reflection
spectroscopy? There are three main advantages, which are
discussed in the following section. Before proceeding with this,
however, we complete the present section by presenting an
expression for ΔL for a film, of thickness much less than the
wavelength of light in the film medium on a highly reflecting
substrate (3),

[13] $\Delta L = (i4\pi\Delta\gamma_n^{(a)}/\lambda_o)\cos\phi_a\varepsilon_a^{1/2}/(\cos^2\phi_a - \varepsilon_a/\varepsilon_b)$

where ϕ_a is the angle of incidence for external reflection, and $\varepsilon = \tilde{n}^2$ is the dielectric constant, with the subscripts a and b standing for ambient and bulk substrate phases respectively. The optical change in the interfacial region upon adsorption is represented by $\Delta\gamma_n^{(a)}$, which is 4π times the change in the complex polarizability per unit area of the interfacial region brought about by adsorption. The subscript n identifies the normal component of this polarizability change (recall E parallel to the surface is essentially zero), while the superscript (a) denotes that the change is to be interpreted as a change in the ambient phase in the interfacial region. One can relate $\gamma_n^{(a)}$ to the polarizibilities of the adsorbed species, given their distribution, through an equation which is the two-dimensional equivalent of the Lorentz-Lorenz equation, or alternatively to the thickness d, and dielectric constant ε_n of a hypothetical surface phase (3) (specifically $\gamma_n^{(a)} = d(1 - \varepsilon_a/\varepsilon_n)$). Both such relations, however, involve models which may well be incorrect. One can also relate ΔL to a change in the substrate phase, or to changes in both phases. Although additional data are required to resolve this ambiguity, it need not concern us directly here, since we shall derive all our information from ΔL through eqs. [9] to [13], for $A_s \simeq 0$, i.e.

[14] $A_e = (2/2.303)\,\text{Re}(\Delta L) \simeq A_p$.

Thus provided that ε_b and ε_a are featureless in the spectral region of interest, ΔL will exhibit absorption bands characteristic of $\Delta\gamma_n$ and hence of the adsorbed species. Although ε_b will generally satisfy this condition for metals in the IR, ε_a may not if IR absorbing molecules are present in the ambient.

The extent to which absorbing species in the ambient gas phase affect the IRES spectra is a matter of considerable importance, since one would like to be able to monitor adsorbed species on surfaces exposed to reactant gases at pressures comparable to those employed in practical catalytic reactors. At first glance it would appear that the ambient phase should not affect the IRES spectra, since in determining A_e, the s-component of the light acts as a reference beam against which changes in the p-component are measured. Since both components traverse exactly the same path, exact compensation for adsorption in the ambient medium would appear to be achieved. This is not quite true, however, as attested to by the presence of ε_a in the equation for ΔL (eq. [13]). The origin of this effect can be thought of as arising from the contribution of the ambient phase to the interphase, which has an effective optical thickness $\sim\lambda_o/|\tilde{n}_b|$. The contribution can be significant for a case in

which the spectra are determined by subtracting a background
spectrum, obtained for a bare surface in vacuum, from one
obtained with the species under study present both on the surface
and in the ambient gas phase. For such a case, it can be shown
that one atm pressure of the species in the ambient phase will
make a contribution to the difference IRES spectrum somewhat less
than that produced by a monolayer of the same species (3). This
then makes it possible to follow the spectrum of surface species
in the presence of high pressures of absorbing species in the
ambient phase, particularly since the gas phase effect is predictable, quantitatively, from its absorption spectrum in a cell
of known length (4).

    IRES Versus Other Reflection Vibrational Spectro-
scopies. In order to achieve a sensitivity sufficient to
detect absorption due to molecules at submonolayer coverages, some
sort of modulation technique is highly desirable. Two
candidates for modulation are the wavelength and the polarization
state of the incident light. The former has been successfully
applied to single crystal studies by Pritchard and co-workers (5),
while the latter is the basis of the Toronto ellipsometric
spectrometer and of the technique employed by Bradshaw and co-
workers (6) and by Overend and co-workers (7). The two different
techniques achieve comparable sensitivities, which for the C-O
stretching mode of adsorbed carbon monoxide amounts to detection
of less than 0.01 monolayer. Sensitivity, of course, is very much
a function of resolution, scan rate, and surface cleanliness.
    The strength of wavelength modulation spectroscopy is its
relative simplicity, its main disadvantage being its sensitivity
to all absorbing species in the optical path and to any fairly
rapid wavelength dependence of the monochromator output or
detector response.
    By far the greatest advantage of polarization modulation, or
IRES, spectroscopy is its lack of sensitivity to absorbing species
anywhere in the optical system except on the test surface. This
makes it the ideal technique for following catalytic reactions on
specularly reflecting surfaces. Its other main advantages stem
from the fact that both the absorption and dispersion reflectance
spectra (Re(ΔL), and the imaginary part of ΔL, Im(ΔL) respectively)
are available from IRES (though not from simpler forms of
polarization modulation spectroscopy). This has two benefits
associated with it. The first benefit is that one can calculate
the absorption and dispersion spectra of the surface film itself
($\propto$ Im($-\Delta\gamma_n$) and Re($\Delta\gamma_n$) respectively) for typical conditions in
which ε_b is complex valued and $\varepsilon_a/\varepsilon_b$ is not negligible compared
to $\cot^2\phi_a$ (see eq. [13]). Ignoring the contribution of the
factor ($\cot^2\phi_a - \varepsilon_a/\varepsilon_b$) to the reflectance absorbance could cause
difficulties in interpreting subtle features of the spectra (line
shape and frequency shifts) in terms of microscopic models of the
surface layer. The second benefit stems from being able to

determine changes in the optical properties of the surface region
of the substrate phase brought about by chemisorption. In the
IR, these changes will relate primarily to the properties of the
conduction electrons near the interface, so that they will appear
in $\Delta\gamma_n$ as very broad features not detectible by wavelength modulation spectroscopy, but appearing as background shifts in the
IRES absorption and dispersion spectra. As the primary effect of
the substrate is probably to introduce a shift in L which varies
linearly with the surface charge on the metal (8), measurement
of these shifts can provide information on chemisorption which
is distinct from but complementary to that obtained from measurements of work function changes.

In summary, the three advantages of IRES over wavelength
modulation photometric spectroscopy are:
1)   its insenstivity to the absorption properties of the ambient
gas phase;
2)   its greater information content allowing $\Delta\gamma_n$ to be determined from the measured spectra;
3)   its ability, again through greater information content, to
monitor substrate changes with increasing coverage.

Over conventional reflectance photometry, IRES retains
advantages 1) and 2), with 3) being partially lost (the substrate
contribution to A_p only is measurable); however, sensitivities
are a factor of 10 or more lower. Polarization modulation
techniques that determine only A_e, and not the corresponding
dispersion property (6,7), retain advantage 1) over wavelength
modulation photometry,but not advantage 2),and retain
advantage 3) only in part, as above.

High resolution electron energy loss spectroscopy, EELS, is
lacking with respect to all three of the foregoing advantages,
and in addition suffers from poor wavelength resolution, limiting
its usefulness in probing subtle interactions between molecules
on a surface. It has the big advantages over all IR techniques
of ready access to the entire vibrational energy spectrum above
about 60 cm^{-1}, at present higher sensitivity, and when used off-
specular, sensitivity to tangential optical modes, i.e. to
absorption processes related to the component of surface polarizability parallel to the interface.

Laser Raman spectroscopy is not at present sufficiently
sensitive for general use in surface studies, though for certain
metal substrates (Ag, Cu, Au) high enough sensitivities are
obtained. The mechanism underlying the enormously enhanced Raman
scattering for these metal substrates is not established,but from
Moskovits' results appears to involve a plasma resonance effect
(9). When it can be used, it has one major advantage over all
other external reflection vibrational spectroscopies in that it
can probe surfaces lying within very strongly IR absorbing media,
such as electrode surfaces immersed in aqueous electrolytes (10).
It can also provide information on tangential optical modes.
Finally, while it has advantage 1) for those substrates that give

rise to enhanced Raman scattering, it does not share advantages
2) and 3) with IRES.

Advantages of IRES for Studying Heterogeneous Reactions.
Due to its insensitivity to the absorption properties of an
ambient gas phase, IRES is admirably suited to the study of
reactions occurring at the solid-gas interface, and hence to the
study of heterogeneous catalysis. For such studies, one requires
a two-dimensional, specularly reflecting reactor. Such a reactor
can be made by depositing the catalytic dispersion onto a face of
an internally reflecting and IR transparent prism, and determining the IRES spectra via internal reflection. Alternatively,
a two-dimensional supported metal catalyst, for example an SiO_2
supported Ni catalyst, could be prepared by first depositing
a thin SiO_2 layer over a specularly reflecting Ni film, then
depositing finely dispersed Ni on this.

There are, of course, a number of excellent experimental
techniques for investigating catalytic reactions in conventional
three-dimensional catalytic reactors (11). However, two-
dimensional reactors have two major advantages over three-
dimensional reactors: 1) the ease with which the surface
properties can be controlled and characterized, and 2) the wide
temperature range available for study.

Concerning the first advantage, one can as easily study
reactions on a single crystal metal surface as on a planar surface
generated to duplicate the properties of a supported metal
catalyst, as described above. Furthermore, with planar surfaces,
LEED, Auger and photoelectron spectroscopies, along with many
other analytical methods of surface science, can provide
characterization of the surface composition and structure.

Concerning the second advantage, reactant species can be
adsorbed onto a two-dimensional reactor at temperatures that
range from close to that for liquid He up to the substrate melt-
ing point. It is not generally possible to dose a three-
dimensional reactor at low temperatures, since the vapour
pressure of one or more of the reactants will commonly be too low
to permit it to diffuse throughout the reactor. In contrast,
even at low temperatures, a two-dimensional reactor is readily
and controllably dosed using effusive molecular beam dosers. An
example of the importance of this advantage of two-dimensional
over three-dimensional reactors is found in a recent study on the
decomposition of methanol on Ni(100) using IRES supported by
other techniques (12). The crystal was dosed with methanol at
170 K or below, then repeated IRES spectra obtained while warming the crystal. At ∿230 K, an intermediate, carbonyl containing
species began forming, which in turn began to transform to
adsorbed CO at ∿260 K. The ability to dose at low temperatures
makes it possible to stabilize reactive intermediates, and hence
to obtain important mechanistic information. In the above

example, the intermediate has been tentatively identified as a
formal moiety.

As a final example of the use of IRES for investigating
heterogeneous catalytic reactions, we mention a preliminary
study of the reaction of CO and H_2 on Ni(110) (2) in which CO
on the surface was monitored in the presence of CO and H_2 in the
gas phase at pressures $\sim$1 atm, while the temperature was raised
in steps. The C-O stretching band for surface species was found
to vanish at temperatures well below the desorption temperature
for CO on Ni(110), the precise temperature being a function of
the ambient pressure and composition. This study, of course,
makes use of the ability of IRES to function with moderately
high pressures of absorbing ambient gases.

It seems to us that the potential of IRES for studies along
the lines of those outlined is considerable, and should be
exploited.

IR Ellipsometer Design. Although many configurations
have been used for ellipsometers operating in the visible region
of the spectrum, the Toronto instrument is the only wavelength
scanning IR ellipsometer with submonolayer sensitivity known to
the authors. This instrument, which was used for the current
studies, has been described in detail elsewhere (13). It consists essentially of two fixed polarizers which bracket a
rotating polarizer and the test surface, with a glow bar light
source at one end, and an IR detector at the other end. The
monochromator may be placed either following the light source,
or preceding the detector. For any of these configurations, and
for perfect linear polarizers, the light intensity reaching the
detector has the following time dependence for a given wavelength
(14),

[15] $I \propto a_o + a_2 \cos(2\omega t - \gamma_2) + a_4 \cos(4\omega t - \gamma_4)$,

where ω is the angular velocity of the rotating polarizer, and
a_o, a_2, a_4, γ_2, γ_4 constants determined by the various optical
parameters. Although general expressions for these constants
have been derived, here we shall give equations for ψ and Δ
(and hence for L, see eq. [10]) only for the special case in
which the stationary polarizer next to the rotating polarizer is
aligned with its direction of polarization parallel to the plane
of incidence, with the other stationary polarizer aligned at
± 45° with respect to this plane, the positive sign indicating
angular displacement in the direction of rotation of the
rotating polarizer:

[16] $\tan\psi = [\tan\gamma_4/(\tan\gamma_4 - 2\tan\gamma_2)]^{1/2}$,

[17] $\cos\Delta = \overline{+}\ \tan\gamma_2\tan\psi$

All of the data presented herein were obtained using these settings and for a mean angle of incidence of 80°. The instrument uses level crossing detectors and digital timers to determine the signal phase angles, γ_2 and γ_4 to an accuracy of about .002°, corresponding in turn to a sensitivity of somewhat better than 10^{-2}% absorption. Various other experimental factors (e.g. fluctuations in the substrate temperature) can easily degrade this figure by up to a factor of five.

Experimental Procedures

Apparatus. The UHV chamber and associated equipment has been described elsewhere (12). The main chamber, and most of the peripherals, were supplied by Vacuum Generators. It is 45 cm in diameter and equipped with effusive molecular beam dosers, LEED optics, LEED and Auger electronics, ion gauge, argon ion gun, and quadrupole mass spectrometer (see Figure 1). A pair of diametrically opposed 20 cm flanges are fitted with 3.75 cm aperture calcium fluoride, IR transmitting, windows obtained from Harshaw Chemical. These were mounted to subtend an angle of incidence of 80° at the crystal surface.

Following baking at 470 K, usually overnight, pressures below 10^{-11} torr were achieved routinely.

Materials. High purity Ar (99.995%), ^{13}CO (99.99%) were obtained in 1ℓ breakneck seal flasks from Scientific Gas Products Ltd. and used without further purification. The ^{13}CO (99.7 atom %) was obtained from Prochem Ltd. in 100 ml breakneck seal flask. The argon for ion bombardment was introduced into the chamber via a leak valve used exclusively for this purpose.

The nickel single crystal (MARZ grade 99.995%, supplied by MRC) was metallographically polished down to 1 μm diamond paste, then chemically polished as described by Graham and Cohen (15). Transmission x-ray diffraction measurements confirmed the (100) direction of the face to $\sim$0.5°. The specimen, of area $\sim$2 cm^2, was spot-welded to a nickel rod fastened to the specimen manipulator. It could be heated by electron bombardment or radiativity and cooled to $\sim$100 K through the nickel rod and a LN$_2$-cooled copper braid. A control unit held the crystal temperature (via a chromel-alumel thermocouple spot-welded to the back of the crystal) constant to within 1 K.

Final cleaning involved repeated argon ion bombardment (5 x 10^{-6} torr, 0.3 μA, 450 V) of the crystal held at 750 K, followed by an anneal for 10 to 20 min at 1100 K, until sharp (1x1) LEED patterns were obtained, and thermal desorption results for CO agreed with those reported by Benziger and Madix (16).

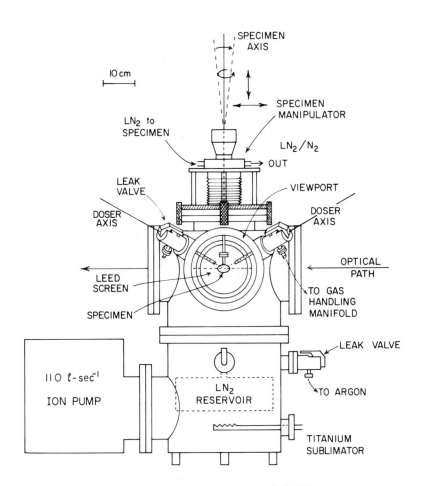

Figure 1.   Schematic of UHV system, equipped with LEED, Auger, mass spectrometer, and IR transmitting windows for IRES

<u>Spectrometer</u>. All of the spectra reported here were
obtained using a LN -cooled InSb detector and a grating and a
slit width giving a resolution $\sim$6 cm^{-1} at 2,000 cm^{-1}. Two or
more scans were made for the "clean" surface and the value for L
from the first subtracted from subsequent scans to yield the
difference spectra, ΔL, which were subjected to a sliding polynomial smoothing procedure designed so as not to degrade the
resolution. The data are reported as percent absorption
($\approx$200 ReΔL, see eqs. [11], [12], and [14]) being 100 times the
fractional decrease in the surface reflectivity for p-polarized
light, brought about by the adsorbed species, less the same
quantity for s-polarized light, which however is essentially zero.

Results and Discussion

<u>CO on Ni(100) at Room Temperature</u>. Ellipsometric
absorption spectra for the indicated effective exposures of CO
on Ni(100) at 298 K are shown in Figure 2. The pressure during
dosing was measured by a nude Bayard-Alpert ionization gauge
located 20 cm from the crystal. As the crystal was in direct line
of sight of the gas doser, the measured pressure was much lower
than that near the crystal, making the measured exposures relative
only. The effective exposures given in Figure 2 and the subsequent
figures were obtained from the measured exposures by multiplying
by 10^2, as this leads to saturation coverage with CO at room temperature for an effective exposure of about 1 L. Since the
sticking coefficient for CO on nickel is roughly unity, such a
conversion leads to reasonable dosage values.

Relative surface coverages were estimated by assuming that
the saturation value for CO on Ni(100) at 298 K is 0.69 monolayers (<u>17</u>), and that infrared band areas are proportional to
coverage. Relative coverages determined in this way were
generally in agreement with those determined by thermal desorption, but in the case of CO rapidly deposited at low temperatures,
the relative coverages (<u>18</u>) determined from IRES fell well below
those determined from thermal desorption. This discrepancy could
stem from significant desorption from the crystal supports,
problems in determining the background spectral level when
dealing with broad IR bands, or a failure in the assumed linear
relationship between CO coverage and total IR band area.

At 298 K, carbon monoxide adsorption is accomplished by the
appearance of a single adsorption band at 2068 cm^{-1} which shifts
to higher frequency with increasing coverage. No other prominent
features appear, although a broad band with some internal
structure spanning 1850 to 1950 cm^{-1} and centred at 1900 cm^{-1} is
clearly present. In this paper, we shall deal only with those
features of this broad band that become prominent at higher
coverages.

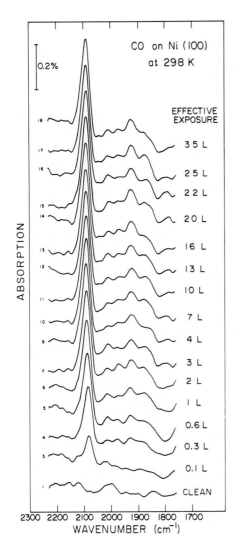

Figure 2. *IRES absorption spectra for CO on Ni(100) dosed as indicated at 298 K*

The 2068 cm^{-1} band is assigned to linearly bonded C-O as it falls within the range observed for gaseous Ni(CO)$_4$ (19-21). It is well known that as the coordination number of CO to metal centres increases, the C-O stretching frequency decreases (22). Thus the broad band centred at ~1900 cm^{-1} is probably due to CO in higher coordination sites.

CO on Ni(100) at Low Temperatures. The data of Figure 3 are typical of those obtained for dosing at low temperatures. As the dose is increased, the spectra develop very much as for the room temperature case (Figure 2), up to an effective dose ~1 L. At an effective dose of ~2 L, however, the low temperature spectrum is very different from that for the room temperature run. In particular, for dosing at 180 K, the broad feature centred at ~1900 cm^{-1} becomes dominated by two fairly well-defined bands at ~1970 and ~1930 cm^{-1}. From data obtained at other temperatures, not reported here, as these two bands begin to grow out of the background with increasing coverage, the 1930 cm^{-1} band is initially the larger of the two. At higher coverages, however, the 1970 cm^{-1} band ultimately becomes the only discernible feature left in the 1850 to 1950 cm^{-1} region, a feature which now dominates the entire C-O stretching region. We assign both these bands to CO in two-fold bridge sites of the Ni(100) surface. These assignments are consistent with the electron loss results of Andersson (21,23) for CO on Ni(100). Andersson reported loss peaks at 44.5 and 81.5 meV (359 and 657 cm^{-1} respectively) accompanying the appearance of a loss peak at 239.5 meV (1931 cm^{-1}). These were assigned respectively to the symmetric and asymmetric Ni-C stretch and the C-O stretch for CO in two-fold bridge sites. For CO on Ni(100) at 298 K the electron loss data revealed a single low energy loss peak at 59.5 meV (480 cm^{-1}) accompanying the loss peak at 256.5 meV (2068 cm^{-1}). These features were assigned to the Ni-C and C-O stretch for linearly bonded CO.

In addition to the development of the very intense band attributed to bridge bonded CO, at ~1970 cm^{-1}, progressive dosing at 180 K leads to pronounced changes in the band attributed to linearly bonded CO, at ~2068 cm^{-1}. Up to ~.9 L, this band shifts in frequency from ~2068 to 2080 cm^{-1}, the same as for dosing at room temperature, but at 2 L, the band broadens somewhat, while at 5 L it is split into bands at 2080 and 2115 cm^{-1}. At doses above 10L, this latter band is the stronger one, with that at 2080 cm^{-1} only weakly present. Finally, when the crystal is equilibrated with 10^{-7} torr of CO, a third band appears in the linear CO region at ~2154 cm^{-1}, vanishing when the system is evacuated once more. As we shall see shortly, the structure giving rise to the 2154 cm^{-1} band is stable under vacuum conditions at still lower temperatures.

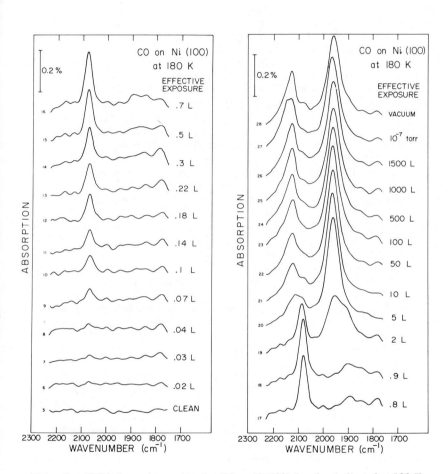

Figure 3.   IRES absorption spectra for CO on Ni(100) dosed as indicated at 180 K

Warm-Up Experiments. The evolution of the spectrum on going from high to low coverages can be followed by saturating the surface at a given (low) temperature, then warming in stages and recording spectra at each temperature. Figure 4 shows the result for such an experiment, in which the Ni crystal, saturated at 180 K in the process of producing the data for Figure 3, was subsequently heated in 15 K steps up to 285 K. Although the spectral changes upon progressive desorption of CO do not follow precisely the same pattern in reverse as that followed during progressive adsorption, they are qualitatively the same. Thus beginning with mainly the 2115 and 1970 cm^{-1} bands at 180 K, by 240 K, the ∿2080 cm band is apparent as a large shoulder on the 2115 cm^{-1} band, while at 270 K, of the two only the 2080 cm^{-1} band remains. Similarly the replacement of the 1970 cm^{-1} band, first by one at 1930, but ultimately by the broad structured band at ∿1900 cm^{-1} can be seen over the range 240 to 285 K. The spectrum obtained on heating to 285 K is virtually identical to that obtained in the dosing sequence of Figure 3 at 0.9 L dose.

The spectra corresponding to saturation coverages at low temperatures were found to be dependent on the details of the adsorption procedure. Thus in contrast to the slow stepwise dosing of Figure 3, leading to the desorption data of Figure 4, the data of Figure 5 were obtained following a single dosing at an effective pressure of 10^{-6} torr (pressure reading of 10^{-8} torr) for 100 S. Even though the data of Figure 5 are for a higher saturation temperature (200 K) than that for the data of Figure 4 (180 K), the spectrum at 200 K shows all three bands near 2100 cm^{-1}.

When the two sets of spectra are compared for the same temperature, differences appear, particularly in the shape of the band structure ∿1900 cm^{-1}. Not until a temperature of 285 K is reached do the two sets give rise to essentially the same spectrum.

The data which resolve most clearly the three bands near 2100 cm^{-1} are presented in Figure 6, and are for ^{13}CO dosed to saturation rapidly, at 125 K. The bands at 2154, 2115, and 2080 cm^{-1} for ^{12}CO are shifted to 2100, 2045, and 2010 for ^{13}CO. The richness of these spectra presumably reflect a corresponding richness in the short range order of CO molecules on the surface achieved at these high coverages.

Sequential Dosing of ^{12}CO and ^{13}CO. Since the linearly bonded sites are occupied first on dosing progressively, up to a coverage of 0.5 (c(2x2) LEED pattern), it should be possible to fill these sites with one isotopic species of CO, then dose further using the other isotopic species to gain information on the filling process beyond 0.5 coverage. The results for such an experiment are shown in Figure 7. The

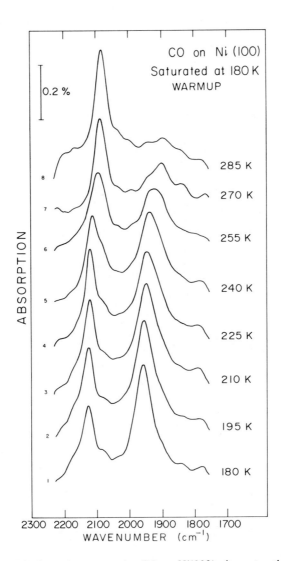

Figure 4.   IRES absorption spectra for CO on Ni(100) after saturating at 180 K (during which the data of Figure 3 were generated), followed by stepwise warming to 285 K

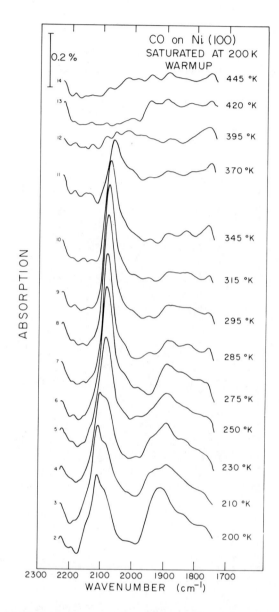

Figure 5.   IRES absorption spectra for CO on Ni(100) after saturating at 200 K,
followed by stepwise warming to 445 K

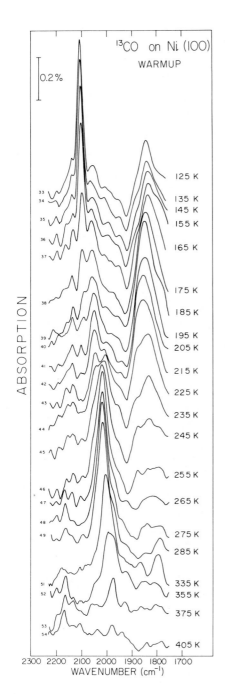

Figure 6.   IRES absorption spectra for ^{13}CO on Ni(100) after saturating at 125 K, followed by stepwise warming to 405 K

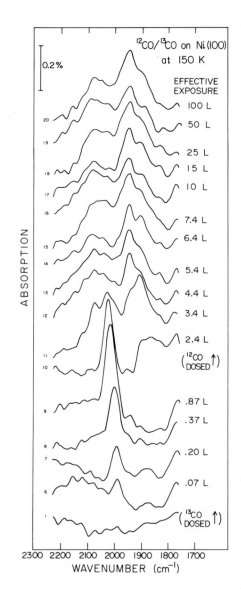

Figure 7. IRES absorption spectra for sequential dosing of ^{13}CO, then ^{12}CO, on Ni(100) at 150 K

Ni(100) surface at 150 K was exposed to ^{13}CO to saturate the linear sites (scans 1-9), then using a second leak valve and gas handling line ^{12}CO was dosed to fill bridged sites (scans 10-20). During dosing of ^{13}CO a single band at 1980 cm^{-1} shifting to 2010 cm^{-1} is observed for the coverage range $0 < 0 < .5$. Upon dosing ^{12}CO initially a broad band at about 1840 cm^{-1} is observed characteristic of bridged ^{13}CO. In addition the 2010 cm^{-1} band due to linearly bonded ^{13}CO diminishes (scan 10). This suggests ^{12}CO has displaced ^{13}CO from linear to bridged sites. With increasing exposure to ^{12}CO, bands at 2080, 2010, 1950, and 1884 cm^{-1} corresponding to ^{12}CO and ^{13}CO in their linear and bridged forms appear. Because the development of a ^{12}CO bridged feature at $\sim$1950 cm^{-1} postcedes that of the 2080 cm^{-1} linear ^{12}CO band, it is suggested that bridged ^{12}CO is formed predominantly by displacement of linear ^{12}CO.

Summary and Conclusions

Our IR studies reveal that bridge-bonded and linearly-bonded CO exist at coverages above that corresponding to the maximum intensity c(2x2) LEED pattern, in agreement with the EELS data of Andersson ([21,23]). Changes in the infrared band profiles, including splitting of the linearly bonded CO peak concurrent with the appearance of bridge-bonded CO, suggests that interactions between adsorbed CO molecules give rise to CO moieties which are distinguishable via their IR bands, due most likely to differences in the short range order about the distinguishable moieties.

Sequential dosing with ^{13}CO and ^{12}CO indicate that above 0.5 coverage (c(2x2))CO probably enters vacant linear sites, displacing neighbouring CO molecules into two-fold bridge sites.

For low temperature dosing, the area of the band attributed to bridge-bonded CO grows to more than twice that ever achieved by the band attributed to linearly-bonded CO for dosing at any temperature, once more in qualitative agreement with Andersson's EELS data. We may interpret this result in one of three ways: (1) The oscillator strength (dipole derivative) associated with the C-O stretching mode for bridge-bonded CO is of the order of 7 times that for the linearly-bonded CO, giving a maximum bridge-bonded CO concentration about 4.0 times that achieved by the linear CO at c(2x2) coverage, and hence presumably a maximum coverage of $\sim$0.7, in agreement with that inferred by Tracy ([25]). (2) The concentration of CO corresponding to the maximum in intensity of the c(2x2) LEED pattern is substantially less than 0.5, again leading to saturation coverages <1. (3) Coverages well in excess of one CO per surface Ni atom can be achieved at low temperatures, contrary to the conclusion inferred by Tracy ([25]) from his results near and above room temperature.

Conclusion (1) is contrary to all previous information

concerning C-O stretch band intensities for carbonyls and adsorbed
CO (26). We can neither rule out (2),nor think of a model that
would lead to such behaviour. Conclusion (3) is contrary to
current wisdom, but the definitive experiments have not been
done.   For the present the matter is unresolved.

Our IR studies provide information on short range surface
structures that is inaccessible by LEED. As many catalytic
processes take place at high pressures, hence frequently high
surface coverages, structural information in the high coverage
domain is clearly valuable in further developing our understanding
of catalytic reactions.

At this point we can only guess at the surface structure
that gives rise to the three different bands for linearly-bonded
CO. IR band areas suggest, however, that the two highest
frequency bands do not appear until CO coverages exceed one monolayer (1 CO per surface Ni atom). They might arise, therefore,
from linearly-bonded CO moieties that share a Ni atom with one or
more bridge-bonded CO.

A more complete analysis of the data presented here will be
given in subsequent publications, along with additional data.

Acknowledgements: The authors are pleased to acknowledge the
support of this research by the Natural Sciences and Engineering
Research Council of Canada and by a special research grant from
the University of Toronto.

Abstract

The potential of IR ellipsometric spectroscopy (IRES) for
investigating surface processes and reactions relevant to gas-
solid heterogeneous catalysis is examined, both for single crystal and model dispersed catalytic systems. With it, structural
and chemical changes can be followed over a wide range of
temperature and gas pressure, allowing one to thermally stabilize
intermediates for investigation, and study surface species under
conditions close to those in practical catalytic reactions.

Data are presented for ^{12}CO and ^{13}CO on Ni(100) which
illustrate the potential of IRES for distinguishing CO in a
variety of surface structures. In addition to bands attributed
to bridge-bonded CO, three distinct bands, attributed to linearly
bonded CO, are observed. The two high frequency bands attributed
to linear CO appear only at high coverages. They are not due to
multilayer adsorption, however, but rather to interactions between
linear- and bridge-bonded CO within a single layer.

Sequential dosing with ^{13}CO and ^{12}CO gave spectra indicating
that above c(2x2) coverage (0.5 monolayer), CO probably enters
vacant linear sites, displacing neighbouring linearly-bonded CO
into two-fold bridge sites.

Literature Cited

1. Yates, Jr., J. T., Chem. Eng. News, 1974, 52. 19.
2. Mahaffy, P. R. and Dignam, M. J., submitted to J. Phys.Chem.
3. Dignam, M. J. and Fedyk, J. D., Appl. Spec. Rev., 1978, 14(2), 249.
4. Dignam, M. J., Rao, B., Moskovits, M. and Stobie, R. W., Can. J. Chem., 1971, 49, 1115.
5. Horn, K. and Pritchard, J., Surf. Sci., 1975, 52, 437.
6. Hoffmann,F. and Bradshaw, A. M., Surf. Sci., 1977, 72, 513.
7. Golden, W., Dunn, D. S., and Overend, J., J. Phys. Chem., 1978, 82, 843.
8. Dignam, M. J. and Fedyk, J. D., J. de Physique, Colloque C5, 1977, C5-C57.
9. Moskovits, M., Solid State Communications, 1979, 32, 59.
10. Jeanmaire, D. L., and Van Duyne, R. P., J. Electroanal. Chem. 1975, 66, 248.
11. see e.g. Robertson, A.J.B., "Catalysis of gas reactions by metals", Logos Press, London, 1970.
12. Baudais, F., Borschke, A., Fedyk, J. D., and Dignam, M. J., Surf. Sci., in press.
13. Stobie, R. W., Rao, B., and Dignam, M. J., Appl. Opt., 1975, 14, 999.
14. Stobie, R. W., Rao, B., and Dignam, M. J., 1975, 65, 25.
15. Graham, M. J. and Cohen, M., J. Electrochem. Soc., 1972, 119, 879.
16. Benziger, J. B. and Madix, R. J., Surf. Sci., 1979, 79, 394.
17. Klier, K., Zettlemoyer, A. C., and Leidheiser, Jr., H., J. Chem. Phys., 1970, 52, 589.
18. Fedyk, J. D. and Dignam, M. J., paper in preparation.
19. Jones, L. H., McDowell, R. S., and Goldblatt, M., J. Chem. Phys., 1968, 48, 2663.
20. Jones, L. H., McDowell, R. S., and Goldblatt, Inorg. Chem., 1969, 2349.
21. Andersson, S., Solid State Commun., 1977, 21, 75.
22. Moskovits, M. and Hulse, J. E., Surf. Sci., 1978, 78, 397.
23. Andersson, S., Solid State Commun., 1976, 20, 229.
24. Ewing, G., J. Chem. Phys., 1962, 37, 2250.
25. Tracy, J. C., J. Chem. Phys., 1972, 56, 2736.
26. Davenport, J. W., personal communication, manuscript submitted to Chem. Phys. Letters, 1980.

RECEIVED June 3, 1980.

Surface Electromagnetic Wave Spectroscopy

R. J. BELL, R. W. ALEXANDER, JR., and C. A. WARD

Physics Department, University of Missouri–Rolla, Rolla, MO 65401

Surface electromagnetic waves or surface polaritons have recently received considerable attention. One of the results has been a number of review articles[1-8], and thus no attempt is made here to present a comprehensive review. These review articles have been concerned with the surface waves, per se, and our interest is in the use of surface electromagnetic waves to determine the vibrational or electronic spectrum of molecules at a surface or interface. Only methods using optical excitation of surface electromagnetic waves will be considered. Such methods have been the only ones used for the studies of interest here.

The first section begins with description of surface electromagnetic waves at a planar, metal-vacuum interface with the goal of understanding why surface electromagnetic waves may be a useful spectroscopic tool. Incidently, metals are the most widely useful class of materials which support surface electromagnetic waves. Then the effects of adding a thin layer of sorbed molecules to the metal surface are outlined. With these results in hand, we turn to a consideration of the three most commonly used optical methods for exciting surface electromagnetic waves and give some examples of studies made using each technique.

Two Media

Surface electromagnetic waves (SEW) on a metal-vacuum interface (often called surface plasmons) are discussed to demonstrate the essential features of SEW. SEW are surface waves in the sense that the electric and magnetic fields decay exponentially as one moves away from the surface, either into the metal or into the vacuum. Figure 1 shows the coordinate system we shall use. The metal-vacuum interface is the $z = 0$ plane, and the metal occupies the $z < 0$ half-space. The direction of propagation is the positive x-direction. The metal has a

0-8412-0585-X/80/47-137-099$05.00/0
© 1980 American Chemical Society

complex dielectric constant $\varepsilon(\omega) = \varepsilon_1(\omega) + i\ \varepsilon_2(\omega)$. From Maxwell's equations and the usual boundary conditions on the electromagnetic field, we find that the only surface wave solutions are analogues of plane waves with electric field components in the x- and z-directions and only one magnetic field component, which is in the y-direction. The electric field has the form

$$\vec{E}_b = E_0(1,0,ik_x/k_{bz})\ \exp\ (ik_x x - k_{bz}z - i\omega t) \tag{1}$$

in the vacuum, and

$$\vec{E}_a = E_0(1,0, - ik_x/k_{az})\ \exp\ (ik_x x + k_{az}z - i\omega t) \tag{2}$$

in the metal. The propagation vector, k, has two (complex) components, k_x and k_z. The subscript a denotes the metal, the subscript b, the vacuum. The boundary conditions require k_x to be the same in both media. Figure 2 illustrates the electric field described by these equations. This is just the electric field pattern that would result from a sinusoidal charge distribution moving along the surface. Because the net charge on the surface is zero, the electric field falls to zero as we move away from the surface.

We also find that the wavelength of the SEW is shorter than the wavelength of a photon of the same frequency traveling in vacuum. It is this property which keeps us from directly exciting SEW by merely shining light on the surface. It is more convenient to consider not λ, but $k = 2\pi/\lambda$. We then find that the x-component of k is

$$k_x(\omega) = \frac{\omega}{c} \left[\frac{\varepsilon(\omega)}{\varepsilon(\omega) + 1}\right]^{1/2} = k_{1x} + ik_{2x}. \tag{3}$$

This equation is called the SEW dispersion curve. For ordinary electromagnetic waves in vacuum, the dispersion curve is $k = \omega/c$. We also find, in order to have surface waves, that

$$\varepsilon_1(\omega) < - 1. \tag{4}$$

Figure 3 shows the dispersion curve plotted for SEW on a gold surface. The frequency for which $\varepsilon(\omega) = -1$ is called the surface plasmon frequency, ω_{sp}, for gold. SEW exist at all frequencies below the characteristic ω_{sp} for each metal. This frequency is in the visible or uv for most metals. For wavelengths longer than about 2 μm, $\varepsilon(\omega) << -1$ and the real part of $k_x(\equiv k_{1x})$ is nearly equal to ω/c. The imaginary part of $k_x(\equiv k_{2x})$ tells how fast the SEW decays as it propagates along the interface. Figure 4 illustrates this by plotting the distance in which the surface wave decreases 1/e in intensity on a copper

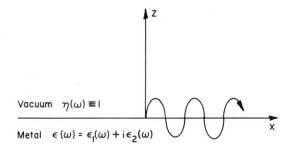

Figure 1.   Coordinate system for surface electromagnetic waves on a metal–vacuum interface

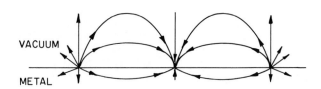

Figure 2.   Schematic of the electric field associated with a surface electromagnetic wave

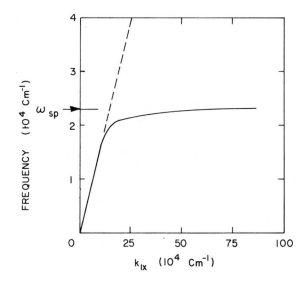

Figure 3.   Dispersion curve for a surface electromagnetic wave on a gold vacuum interface

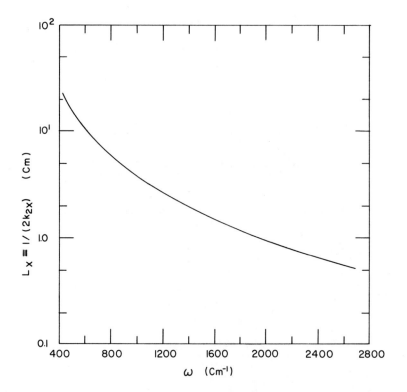

Figure 4. Propagation distance, L_x, for surface electromagnetic waves on a Cu–vacuum interface

substrate. Despite the fact that metals are highly absorbing, it can be seen that SEW propagate several cm at a wavelength of 10μm. This is because most of the energy in the SEW is in the vacuum above the metal.

The real parts of k_{az} and k_{bz} tell us how far the electric and magnetic fields penetrate into the metal (a) and vacuum (b), respectively. Figure 5 shows the distance in which these fields fall to 1/e of their value at the surface for a vacuum-copper interface. The fields extend a much greater distance into the vacuum than they do into the metal.

We note that ionic crystals may have dielectric functions satisfying Eq. (4) for frequencies between their transverse and longitudinal optic phonon frequencies. SEW on such crystals are often called surface phonons or surface polaritons and the frequency range is the far IR.

Three Media

We model a metal with sorbed molecules as shown in Fig. 6. The molecules form a thin absorbing layer of thickness d, with a complex dielectric constant $\eta(\omega) = \eta_1(\omega) + i\eta_2(\omega)$. The analysis of this system is rather complicated, and the reader is referred to the paper by Bell, et al. for details of the exact solution and various approximations.[9] For thin films, Bell has obtained a reasonably simple approximation for the absorption of an SEW due to the overlying film. If the SEW are propagated a distance D on the metal substrate, with and without the thin film, the ratio of the SEW intensities is given by

$$T(\omega,D) \equiv I(\omega,D)/I_0(\omega,D) = \exp(-2\pi\omega\Delta_2 D). \qquad (5)$$

$I(\omega,D)$ is the SEW intensity with the film, and $I_0(\omega,D)$ is the intensity without. The parameter Δ_2 is related to the dielectric functions of the film and the metal substrate by[9]

$$\Delta_2 \approx \frac{4\pi\omega d}{(-\varepsilon_1)^{1/2}} \frac{(\eta_1^2 + \eta_2^2 - \eta_1)\varepsilon_2/2(-\varepsilon_1) + \eta_2}{(\eta_1 + \eta_2\varepsilon_2/2\varepsilon_1)^2 + (\eta_2 + \eta_1\varepsilon_2/2\varepsilon_1)^2} \qquad (6)$$

Figure 7 shows a plot of T versus ω for a monolayer and a five layer thick film of Cu_2O on copper in the region of the strong lattice absorption of Cu_2O.[9] Two features should be noted. The presence of a monolayer is easily seen, and a decrease in transmission occurs even when the Cu_2O film is nonabsorbing. The decrease in transmission in regions where the film is non-absorbing allows the determination of the thickness of the film from the measured transmission using[10]

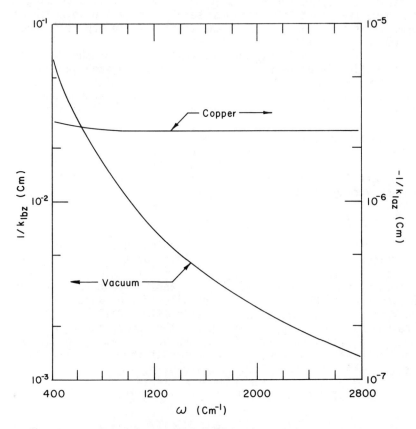

Figure 5. Decay lengths for the electromagnetic field of a surface electromagnetic wave on a Cu–vacuum interface. The right-hand scale is for the decay length into the Cu and the left-hand scale for the decay length into vacuum.

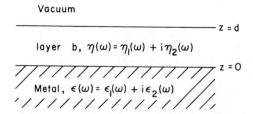

Figure 6. Schematic of thin layer of thickness d on a metal substrate

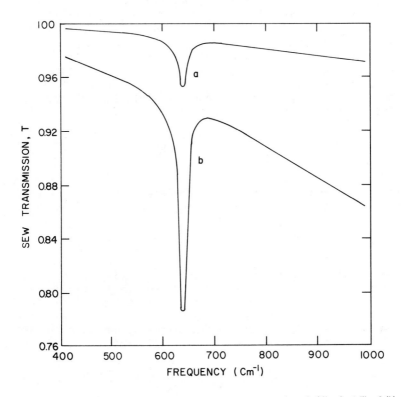

Journal of Non-Crystalline Solids

Figure 7. The surface electromagnetic wave transmission between two prisms separated by 10 cm for (a) one monolayer of Cu₂O on Cu and (b) five monolayers of Cu₂O on Cu (29)

$$d = \frac{\sqrt{-\varepsilon_1}}{8\pi^2\omega^2 D} \left(\frac{\eta_1}{\eta_1-1}\right) \left(\frac{4\varepsilon_1^2 + \varepsilon_2^2}{-2\varepsilon_1\varepsilon_2}\right) \ln\left(\frac{I_0}{I}\right) \qquad (7)$$

Optical Excitation of Sew

Above we saw that an SEW has a wavelength shorter than an electromagnetic wave of the same frequency and hence cannot be directly excited by an incident electromagnetic wave. Otto[11] and Kretschmann and Raether[12] realized that the technique of attenuated total reflection could be adapted to excite surface waves. An electromagnetic wave traveling in a medium with a refractive index n has a wavelength λ/n where λ is the wavelength in vacuum. In terms of k, this means $k = \eta\omega/c$. If total internal reflection takes place at the surface of this material, then an exponentially decaying electromagnetic field exists beyond the surface, and the component of k along the surface (x-direction) is

$$k_{1x} = \eta(\omega/c) \sin \theta. \qquad (8)$$

The angle of incidence is θ, as shown in Fig. 8. To obtain total internal reflection we must have θ greater than the critical angle for total internal reflection, θ_c, which is given by

$$\sin \theta_c = 1/n$$

By adjusting θ (within the restriction $\theta > \theta_c$), Otto realized he could match k_{1x} of the surface wave [Eq. (3)] to the k_{1x} of the totally internally reflected wave [Eq. (8)] and hence excite the surface wave. Note the gap of thickness t shown in Fig. 8 is necessary to obtain the total internal reflection and that the surface wave is excited by the exponentially decaying fields below the prism. Otto[11] found that when θ satisfied

$$k_0 \ n \sin \theta = k_{1x} = \omega/c \ \mathrm{Re} \left[\frac{\varepsilon(\omega)}{\varepsilon(\omega)+1}\right]^{1/2}, \qquad (9)$$

the total internal reflection was frustrated (no longer was total) because energy was being lost to the excitation of surface waves. The gap is usually air (or vacuum), but low refractive index materials have been used.[12,13] The prism need not be triangular in section; often hemispherical prisms are used to avoid refraction as the beam enters and leaves the prism.

The single prism measurements can also be done using the configuration shown in Fig. 9. Here the thin metal film itself on the prism base serves as the low index gap for exciting SEW on the metal-air interface below. This type of attenuated total reflection excitation of surface waves was first used by

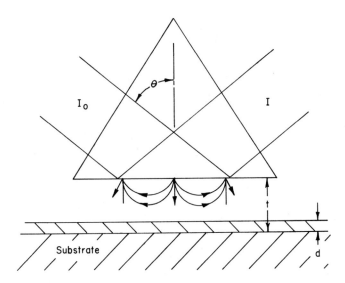

Figure 8. ATR method for exciting surface electromagnetic waves by using a single prism as developed by A. Otto (11). The angle of incidence of the light on the prism base is θ.

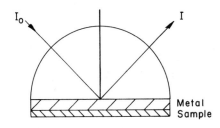

Figure 9. ATR method for exciting surface electromagnetic waves by using a single prism with the metal film serving as a gap as developed by Kretschmann and Raether (12)

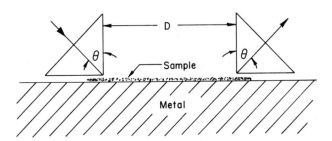

Figure 10. The two prism method for surface electromagnetic wave spectroscopy as developed by Schoenwald, Burstein, and Elson (14)

Kretschmann and Raether.[12] In order to have the needed total internal reflection at the prism-metal interface, one is limited (approximately) to wavelengths shorter than 3 μm. With gold, silver, or copper films and a prism with large refractive index (e.g. Ge), it is possible to reach 5 or 6 μm.

A modification of this single prism technique was suggested by Schoenwald, et al.[14] to study thin layers on a metal substrate. The basic idea was to split the prism in two pieces, the first of which launches a surface wave; the second recovers the wave. This arrangment is sketched in Fig. 10. This technique requires a propagation distance of one mm or longer, which means wavelengths longer than 2 μm. See Fig. 4, for example. Because a highly collimated beam is required for efficient coupling of the incident radiation to the SEW, a laser source is required. This is not true for the single prism method.

Chabal and Sievers[15] developed another coupling scheme for SEW transmission spectroscopy. The arrangement is shown in Fig. 11. The beam is incident upon the prism base at the proper angle for coupling just at the edge of the metal film on the prism base. The SEW is then decoupled at the other edge. Best coupling was obtained with a saw toothed edge on the film. Once again a laser source is required.

We mention that a grating ruled on a metal surface can also be used to excite SEW.[7] Since this method is difficult and has been little used, it will not be considered further.

Results

The single prism technique has the advantage that ordinary sources with a monochromator can be used. The two prism technique requires an extremely well collimated beam, and only lasers are satisfactory as sources. However, the two prism technique is more sensitive to absorption in the thin film. We shall give some examples of data obtained using each method. It must be admitted that much of the work to date has been of an exploratory nature, designed to demonstrate the utility of spectroscopy using surface waves.

The excitation of an SEW using a single prism is seen as a dip in the reflectivity from the prism base (see Fig. 8) when the angle of incidence satisfies Eq. (8). The location in angle of the minimum in reflectivity, the depth of the minimum and its width are very sensitive to the presence of an overlayer on the metal substrate. Pockrand, et al.[16] have pointed out that this technique is considerably more sensitive than ellipsometry for measuring the optical properties of thin films. They also discuss the problems due to the fact that the dye films are really anistropic.

The reason for the increased sensitivity over ellipsometry is that the position of the reflectivity minima is extremely sensitive to the presence of attached molecules. A particularly

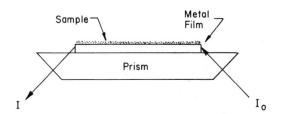

Figure 11.   The edge coupling scheme for surface electromagnetic wave spectroscopy as developed by Chabal and Sievers (15)

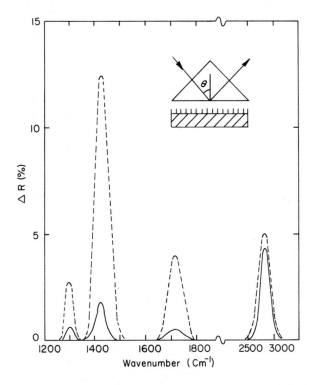

Optics Communication

Figure 12.   The vibrational absorption spectrum for Mn stearate for (———) a two layer film and (– – –) a 10 layer film. These spectra were obtained by using the Otto single prism method as shown in the inset. The Mn stearate molecules lie on the Ag substrate (18).

nice demonstration of this sensitivity was given by Weber.[17] He used the Kretschmann-Raether configuration with a modulated angle of incidence technique which allowed extremely precise location of the reflectivity minima. Weber was able to observe 0.2% of one monolayer of oxygen on copper.

The vibrational spectrum of Mn stearate has been measured using the Otto configuration by Hjortsberg, et al.[18] Figure 12 shows the change in the reflectivity from the prism base due to the addition of two layers and 10 layers of Mn stearate. Clearly monolayers can be easily detected.

A number of dye systems have been studied using the Kretschmann-Raether geometry.[16,19] Although the excitations observed in dyes are electronic, not vibrational, they are mentioned because no vibrational studies using this technique are known to us. Figure 13 shows how the absorption of the dye affects the coupling angle for a range of wavelengths. The change in the depth of the reflectivity is also shown. The dye was S 120(1-octa-decyl-1')-methy-2,2'-cyanine perchlorate) and was one monolayer thick. It has a strong absorption at 575 nm. The solid curves show the calculated values, assuming the bulk values of the dielectric function can be applied to the thin film.

Attenuated total reflection studies using the Kretschman-Rather configuration have also been done in ultrahigh vacuum by Chen and Chen.[20] They use the clever idea shown in Fig. 14 to avoid the need for more than one vacuum window. They observed the change in coupling angle as a monolayer of Cs on a silver film was oxidized. This change in angle is plotted as Fig. 15. All of these data suggest that the vibrational spectrum of molecules sorbed on a clean substrate can be studied.

Another example of the use of SEW spectroscopy to study reaction rates is the measurements of the diffusion of Al into gold with the subsequent formation of Au_2Al. Loisel and Arakawa[13] used a MgF_2 spacer instead of a vacuum spacer, and gold and aluminum were plated upon this spacer, as shown in Fig. 16. The SEW are excited at the MgF_2 - gold interface, and the exponentially decaying field in the gold reaches the gold - Au_2Al interface. As the reaction occurs, the coupling angle changes as the Au-Au_2Al interface migrates into the gold film. The coupling angle versus reaction time at a temperature of 80C is shown for several wavelengths in Fig. 17. From this data, the authors obtained the diffusion coefficient of Al into Au.

Now we turn to two-prism measurements using SEW. Figure 18 shows the measured propagation distance using a CO_2 laser as a source in the region around the 1050 cm^{-1} absorption in cellulose acetate.[10] The film was 15A thick and the prisms were separated by 3 cm. Note that the propagation distance (relative to the copper substrate with no cellulose acetate) is reduced even in regions where the cellulose acetate is transparent. This allows the film thickness to be determined from Eq. (7). The result agrees well with the thickness obtained by ellipsometry.

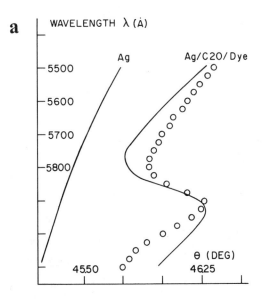

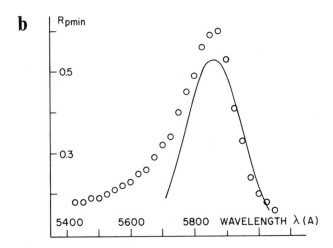

Thin Solid Films

Figure 13. (a) Wavelength vs. coupling angle for S-120 dye on a Ag substrate; (b) the minimum reflectivity vs. wavelength for the same system. The measured points are shown as open circles and the solid curves calculated from bulk values of the dielectric function for S-120 dye. The measurements were made using the Kretschmann–Raether arrangement (19).

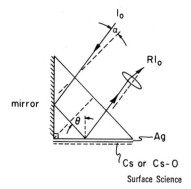

Figure 14.   Experimental arrangement
used by Chen and Chen (20) to observe
the oxidation of a Cs film on a Ag film

Surface Science

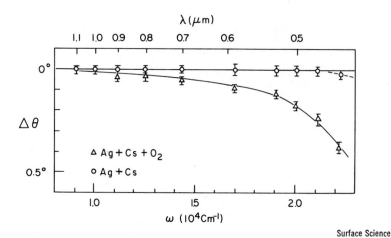

Figure 15.   Change in coupling angle as a function of wavelength for a clean Cs
film and an oxidized film (20)

Surface Science

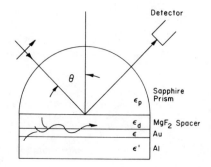

Figure 16.   Hemicylindrical prism and
MgF_2 spacer used by Loisel and Arkawa
(13) to study the diffusion of Al into Au

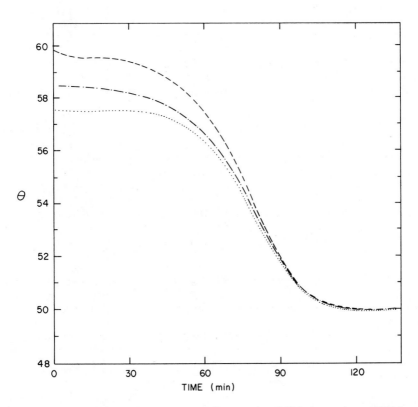

Figure 17.   Coupling angle, θ, vs. time for (− − −) λ = 5900Å, (− · − ·) λ = 6093Å, and (· · ·) λ = 6200Å. The temperature was 80°C (13).

The vibrational spectrum of benzene around 1000 cm^{-1} has also been measured.[10] Benzene was physisorbed on a cooled copper substrate in the vacuum chamber. Figure 19 shows the transmission for several thicknesses of benzene and a prism separation of 3 cm. The thickness was determined from the measured transmission in transparent regions using Eq. (7). The solid curves were calculated from Eqs. (5) and (6) using optical constants for benzene obtained from an ordinary transmission experiment.[21] The benzene film was assumed to be isotropic. Of the two absorption lines seen, one belongs to an in-plane vibrational mode, and one to an out-of-plane vibration. Since the electric field of the SEW is primarily perpendicular to the surface, the benzene molecules are clearly not all parallel or all perpendicular to the copper surface. Also it should be noted that the frequencies are the same (within the experimental resolution) as those of solid benzene[22] and of nearly the same width. These features indicate that the benzene interacts only weakly with the copper surface, as would be expected for physisorbed molecules.

Semiconductors with free carriers can support SEW. This has been demonstrated for a number of materials, but little use has been made of them as substrates for adsorbed molecules. To illustrate that interesting possibilities exist, Fig. 20 shows the propagation distance for SEW on GaAs as a function of free carrier concentration for a frequency of about 80 cm^{-1}. These data were taken with a molecular laser.[23] The observed distances are of order 1 cm, sufficient for use with the two prism technique.

Ionic crystals also support SEW, but again no data exists where they have been used as substrates for attached molecule studies. That such studies may be feasible is illustrated in Fig. 21, which shows measured and calculated propagation distances for $SrTiO_3$ in the far infrared.[24] Again, these measurements were made with a molecular laser as a source. Unfortunately, for many crystals the frequency region over which SEW exist is very narrow (between the transverse and longitudinal optic phonon frequencies), and propagation distances are very short. However, ferroelectrics (and near-ferroelectrics like $SrTiO_3$) may prove useful substrates for SEW spectroscopy.

Conclusion

Surface wave spectroscopy is still in its infancy, and most of the examples we have mentioned have been demonstrations of what can be done. It has clearly been demonstrated that vibrational and electronic spectroscopy can be done on molecules sorbed on metal substrates at less than monolayer coverages. In contrast to many techniques for surface spectroscopy (e.g. ultraviolet photoelectron spectroscopy, Auger electron spectroscopy), SEW spectroscopy does not require ultrahigh vacuum and can be done in the presence of reactant gases. This should make it a valuable tool for the study of catalytic reactions. Certainly

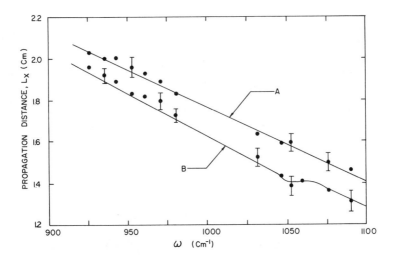

Figure 18. (A) Propagation distance for a Cu–vacuum interface as a function of frequency; (B) propagation distance for Cu with a 15Å layer of cellulose acetate. The curves are theoretical fits calculated from Equations 5 and 6 (10).

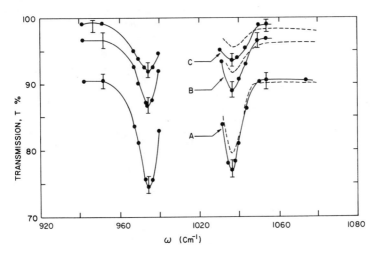

Figure 19. Transmission spectrum for benzene physisorbed on a Cu substrate: (– – –) calculated for (A) 25Å, (B) 10Å, and (C) 5Å thicknesses; (——) drawn through the measured points (10).

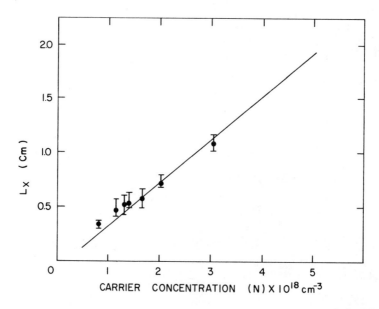

Surface Science

Figure 20. Propagation distance of surface electromagnetic waves on GaAs as a function of carrier concentration for a frequency of 84 cm⁻¹ (23)

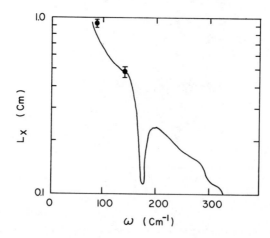

Surface Science

Figure 21. Propagation distance of surface electromagnetic waves on SrTiO₃. The solid curve is calculated from the two points measured with a molecular laser (24).

it is simpler than ellipsometry for monolayer and submonolayer coverages. It is also simpler, and at least in some cases, more sensitive, than high angle of incidence spectroscopy.[25] The two prism method, while the most sensitive, has the disadvantage of requiring a tunable coherent source in the infrared. Such a source was constructed in our laboratory, but has proved complicated to operate and produces marginally sufficient power.[26,27] However, recent developments in Raman lasers suggest that these will provide satisfactory sources.[28]

Acknowledgements

This work supported in part by grants from the Air Force Office of Scientific Research (AFOSR 76-2938) and the National Science Foundation (DMR 75-19153).

Literature Cited

1. Otto, A., in Seraphin, B., Ed. "Optical Properties of Solids: New Developments"; North-Holland: Amsterdam, 1976; p. 677.

2. Mills, D. L. Burstein, E., _Rept. Progr. Phys._, 1974, 37, 817.

3. Otto, A., in Queisser, H. J., Ed. "Festkorperprobleme 14"; Vieweg: Braunschweig, 1974; p. 1.

4. Kliewer, K. L. and Fuchs, R., in Prigogine, I. and Rice, S. A., Eds. "Advances in Chemical Physics", Vol. 27, Wiley; New York, 1974; p. 335.

5. Economou, E. N. and Ngai, K. L., in Prigogine, I. and Rice, S. A., Eds. "Advances in Chemical Physics", Vol. 27, Wiley; New York, 1974; p. 205.

6. Burstein, E., Chen, W. P., Chen, Y. J. and Hartstein, A., J. Vac. Sci. Technol._, 1974, 11, 1004.

7. Ritchie, R. H., _Surface Sci._, 1973, 34, 1.

8. Ruppin, R. and Englman, _Rept. Progr. Phys._, 1970, 33, 101.

9. Bell, R. J., Alexander, R. W. Ward, C. A., Tyler, I. L., _Surface Sci._, 1975, 48, 253.

10. Bhasin, K. Bryan, D., Alexander, R. W., Bell, R. J., _J. Chem. Phys._, 1976, 64, 5019. Eq.(7) of this paper contains a typographical error. The numerator should have an ω_p rather than ωp^2.

11. Otto, A., _Z. Physik_, 1968, 216, 398.

12. Kretschman, E., Raether, H.,<u>Z. Naturforsch</u>, 1968, 23a, 2135.

13. Loisel, B. A., Arkawa, E. T., in press.

14. Schoenwald, J., Burstein, E., Elson, J. M., <u>Solid State Commun.</u>, 1973, 12, 185, and private communication.

15. Chabal, Y. J., Sievers, A. J., <u>Appl. Phys. Lett.</u> 1978, 32, 90.

16. Pockrand, I., Swalen, J. D. Santo, R., Brillante, A., Philpott, M. R., <u>J. Chem. Phys.</u>, 1978, 69, 4001.

17. Weber, W. H., <u>Phys. Rev. Lett.</u> 1977, 39, 153.

18. Hjortsberg, A., Chen, W. P., Burstein, E. Pomerantz, M., <u>Opt. Commun.</u>, 1978, 25, 65.

19. Wahling, G., Raether, H., Mobius, D., <u>Thin Solid Films</u>, 1979, 58, 391.

20. Chen, W. P., Chen, J. M., <u>Surface Sci.</u>, in press.

21. Anderson, G. R., Person, W. B., <u>J. Chem. Phys.</u> 1962, 36, 62.

22. Zwerdling, S., Halford, R. S., <u>J. Chem. Phys.</u>, 1955, 23, 2221.

23. Begley, D. L., Alexander, R. W., Ward, C. A., Bell, R. J., <u>Surface Sci.</u>, 1979, 81, 238.

24. Miller, R., Begley, D. L., Ward, C. A. Alexander, R. W., Bell, R. J., <u>Surface Sci.</u>, 1978, 71; 491.

25. Greenler, R. G., <u>J. Vac. Sci. Technol.</u>, 1975, 12, 1410.

26. Sorokin, P. P., Wynne, J. J., Lankard, J. R., <u>Appl. Phys. Lett.</u>, 1973, 22, 342.

27. Tyler, I. L., Alexander, R. W., Bell, R. J., <u>Appl. Phys. Lett.</u> 1975, 27, 346.

28. Grasiuk, A. A., Zubarev, I. G., <u>Appl. Phys.</u>, 1978, 17, 211.

29. Alexander, R. W., Bell, R. J., <u>J. Non-Cryst. Sol.</u>, 1975, 19, 93.

RECEIVED June 3, 1980.

Raman Spectroscopic Studies of Surface Species

B. A. MORROW

Department of Chemistry, University of Ottawa, Ottawa, Ontario, K1N 9B4, Canada

As this Volume illustrates, surface vibrational data can be obtained using a wide variety of experimental techniques. Raman spectroscopy is a particularly useful method insofar as there are virtually no restrictions as to the type of surface which can be studied (oxides, oxide supported catalysts and bulk metals), the accessible frequency range (50 - 4000 cm^{-1}) or the ambient gas pressure (one atmosphere to UHV). By way of comparison, although transmission infrared spectroscopy is widely used to study the adsorption of gases on high area oxides and oxide supported metals (1,2,3), (i) it cannot be used to study adsorption on single crystal metal surfaces, (ii) it sometimes cannot be used to investigate low frequency modes because most oxides are opaque to infrared radiation in certain spectral regions below $\sim$ 1000 cm^{-1}, and (iii) different and sometimes expensive optical materials are required for a complete spectral analysis and more than one spectrometer might be required. Other techniques can overcome some of these difficulties and can be used over a wide spectral range (electron energy loss, IR reflectance, IR ellipsometry, diffuse reflectance) but there are sometimes restrictions on the type of material which can be studied (either metals or oxides), the ambient gas pressure, and the spectral resolution.
    In contrast, recent work (4-12) has shown that Raman spectroscopy can be used to study (i) adsorption on oxides, oxide supported metals and on bulk metals [including an unusual effect sometimes termed "enhanced Raman scattering" wherein signals of the order of 10^4 - 10^6 more intense than anticipated have been reported for certain molecules adsorbed on silver], (ii) catalytic processes on zeolites, and (iii) the surface properties of supported molybdenum oxide desulfurization catalysts. Further, the technique is unique in its ability to obtain vibrational data for adsorbed species at the water-solid interface. It is to these topics that we will turn our attention. We will mainly confine our discussion to work since 1977 (including unpublished work from our laboratory) because two early reviews (13,14) have covered work before 1974 and two short recent reviews have discussed work up to 1977 (15,16).

0-8412-0585-X/80/47-137-119$05.50/0
© 1980 American Chemical Society

Experimental Considerations

The basic requirement for studying Raman scattering are, (a) a monochromatic light course [usually a laser operating in the visible region], (b) a spectrograph to disperse the components of the scattered light [usually a double or triple monochromator in order to reduce stray light], and (c) a photomultiplier and photon counting detection system, sometimes used with a computer for data manipulation and signal averaging. For bulk solids or liquids the scattered light is generally collected at 90° with respect to angle of incidence of the laser, although sometimes a 180° "back scattering" arrangement is used.

In studying molecules adsorbed on surfaces, a wide variety of laser illumination and light collection geometries have been employed (13-16), the exact geometry sometimes being dictated by the necessity of containing the sample to be studied in a suitable cell for pretreatment. The latter may range from a simple Pyrex tube containing an optical flat (4) to complex UHV chambers which permit use of other surface characterization techniques (5,17,18). Further, in many adsorption studies one must be careful that the laser itself will not induce desorption due to local heating and, in lieu of simply using low laser power levels (20-200 mW), some groups have used cylindrical lens focus techniques in order to spread out the laser beam into a line image on the sample, (8,13-16) or else the sample has been rotated so that the energy flux of a focused laser spot is decreased (9,10,19,20).

Powdered oxides or oxide supported samples have generally been studied as pressed discs or extrudates and we have found (21) that optimum signal intensity has generally been achieved with geometries similar to those illustrated in Figure 1, particularly the near 180° back scattering geometry (Figure 1A) where the laser beam strikes the sample at about 10° from the normal such that the "reflected" and scattered light is directed with near normal incidence into the spectrometer. For bulk metals or evaporated metal films, a 90° illumination-scattering geometry has generally been used (as in Figure 1B) but the angle β has been varied considerably depending on the nature or purpose of the study (22). Greenler and Slager (23) have discussed the question of the optimum illumination-scattering geometry for studying the Raman spectra of adsorbed molecules on metals.

A major problem which is endemic to Raman studies of high surface area oxides is background fluorescence, a problem so severe in some cases that weak signals due to Raman scattering are simply lost in the background (13-16). A typical background spectrum of a silica sample is shown in Figure 2, (this also shows the relatively weak features due to SiO_2 itself below 1000 cm^{-1}) where a very broad maximum near 2500 cm^{-1} is observed using 488.0 nm irradiation (i.e., at 18500 cm^{-1} absolute). This fluorescence maximum, which shifts in terms of Raman displacement but remains at the same absolute frequency if the laser frequency is changed (21), is sometimes greatly diminished if the

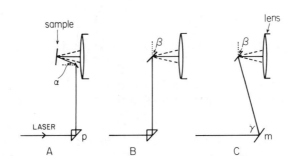

Figure 1. Schematics of various laser illumination geometries. The laser beam strikes either a prism (p) or a mirror (m) and is directed upward toward the sample. The "lens" is the collection lens of the spectrometer (21).

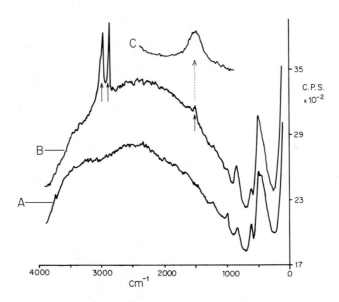

Figure 2. (A) Raman spectrum of silica; (B) Raman spectrum of methylated silica, new bands are marked with vertical arrows; (C) spectrum of the weak feature at 1465 cm⁻¹ with 10-fold wavenumber and ordinate expansion. Laser power 0.6 W for (A) and (B), 1.7 W for (C) (21).

sample is heated in an oxygen atmosphere at 700°C, a result which has been attributed by many to the oxidation of trace quantities of hydrocarbon impurities (4,18,13,14,15,16). With zeolites the presence of trace quantities of ionic iron impurities has also been discussed as a contributing factor (24), and Jeziorowski and Knözinger (25) have presented evidence which indicates that surface hydroxide ions may also be a source of part of the fluorescence. Thus, in studies of oxide supported molybdenum oxide catalysts [see below] part of the elimination of strong fluorescent background has been attributed to removal of OH during catalyst processing (10), although other theories have been advanced in these cases (19). Finally, Saporstein and Rein (26) have recently reported that the fluorescent background from a 4A zeolite can be greatly diminished by washing in 0.2 N NaOH and calcining at 400°C. This should prove useful in future studies of zeolites where, because of possible crystal structure collapse, it may not be desirable to subject the sample to high temperatures.

In the discussion which follows we will frequently make reference to the "quality" of observed Raman spectra. Raman intensities are usually reported in terms of photon counts per second (c.p.s.) and whereas pure liquids frequently exhibit Raman spectra in which the signal to noise ratio is greater than 1000 and maximum peak intensities are in the range $10^3 - 10^6$ c.p.s. (depending on laser power, usually in the range 20-4000 mW, and on the slit width of the spectrometer), the sensitivity of the technique for studying surfaces is often far less than this and wherever possible we have quoted signal intensities for comparative purposes.

Further, in cases where the background fluorescence cannot be entirely eliminated, the Raman spectrum of an adsorbed species appears as a superimposed signal. Figure 2B shows the Raman spectrum of a methylated silica, a sample where all surface SiOH groups have been replaced by $SiOCH_3$ groups. The sharp features near 3000 cm^{-1} due to $\nu(CH)$ modes will be discussed further later, but the signal to fluorescence intensity can be estimated from the ordinate scale.

Adsorption on Zeolites

The adsorption of carbon monoxide on metal catalysts (3) has been the model system of choice for the development of many vibrational spectroscopic techniques (IR, EELS and reflection-absorption) where energy exchange depends on a change in dipole moment during vibration. Pyridine (13-16) has historically been used for the development of the Raman technique for studying surfaces because energy exchange (inelastic scattering) occurs when a molecule undergoes a change in polarizability during vibration. Thus, whereas the CO stretching mode of adsorbed CO yields an intense band in the infrared spectrum, the in-plane symmetric (A_1) ring deformation modes of pyridine have a large Raman scattering cross section. Figure 3A shows a portion of

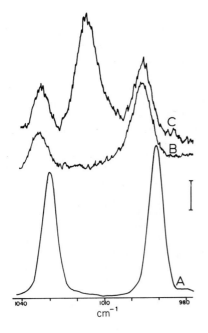

Figure 3.   Raman spectrum of pyridine in the region of the v_1 (991–1016 cm^{-1}) and v_{12} (1030–1036 cm^{-1}) fundamentals: (A) liquid pyridine, (B) pyridine adsorbed on a NaX zeolite, and (C) pyridine adsorbed on a Zn^{++} exchanged (78% exchange) NaX zeolite.   The vertical bar represents: (A) 2000 Hz, (B) 100 Hz, (C) 50 Hz.

the Raman spectrum of liquid pyridine in the region of the particularly intense ν_1 (991 cm^{-1}) and ν_{12} (1030 cm^{-1}) bands. The
symmetric ν_1 ring breathing mode is most useful from the surface
diagnostic point of view because it has been shown to shift in a
predictable way when pyridine interacts with various sites on an
oxide or zeolite surface ([8],[13],[14],[15],[16],[27]). Early studies ([13],
[14],[24],[27]) using model systems and well characterized oxides have
shown that there are fairly clear domains for this mode depending
on whether adsorption occurs via physical adsorption (991 cm^{-1}),
hydrogen-bonding (994-1008 cm^{-1}), proton transfer due to Brønsted
acidity (1009-1012 cm^{-1}) or coordination to electron deficient Al
atoms due to Lewis acidity (1016-1025 cm^{-1}). Accordingly, the
Raman spectra of adsorbed pyridine provides a useful probe for
the determination of the types of adsorption sites which are
present on oxide surfaces, and much of the earlier work has been
reviewed ([13],[14],[15],[16]).

The extension of these studies to zeolites has yielded some
interesting results. In an early study Egerton, Hardin and
Sheppard ([27]) showed, in agreement with Ward's previous infrared
results ([28],[29]), that pyridine adsorbs mainly to the cation in a
series of cation exchanged Y zeolites but that there is a linear
shift of ν_1 to higher frequency with the electrostatic potential
(or charge to radius ratio, e/r, of the exchange cation).

We recently carried out a detailed study ([8]) of the X-zeolite
system (NaX and 10 cation exchanged samples) and found a similar
ν_1 vs. e/r correlation for adsorbed pyridine and for some metal
nitrate-pyridine-nitromethane solutions. Typical spectra in the
ν_1 spectral region showing the attainable signal-to-noise ratio
for adsorption on the NaX zeolite, and on a ZnNaX zeolite for
which 78% of the Na$^+$ had been exchanged with Zn^{++}, are shown in
Figures 3B and 3C respectively. We concluded that adsorption
occurred exclusively via a direct interaction between the pyridine
lone pair electrons and the positively charged cation. Two ν_1
bands were generally observed when there was residual unexchanged
Na$^+$ as can be seen in Figure 3C where the Zn^{++}-pyridine band is at
1016 cm^{-1} and that due to Na$^+$-pyridine is unchanged at 996 cm^{-1}.
We further postulated that where ionic size is not a consideration, the exchange cations (monovalent and bivalent) are distributed at least among sites S_I and S_{II}. In the case of the Cs$^+$
exchanged zeolite some unusual frequency shifts were observed
for other pyridine modes relative to the other exchanged zeolites,
the sample was faintly pink at high pyridine coverage and we
suggested that some photo reduction might have occurred resulting
in the additional formation of a $Cs_x^{(x-1)+}$-py^{+1} species.

Additional recent Raman work has shown that other highly
polarizable molecules can also be used to probe the properties of
zeolites. Freeman and Unland ([7]) studied the adsorption of
benzene on a series of alkali exchanged X and Y zeolites and also
used the measured shift of the ν_1 ring breathing mode (992 cm^{-1}
for liquid benzene) as a diagnostic measure of the interaction
with the exchange cations. The spectra were comparable in quality
to those shown in Figure 3B,C. In all cases the ν_1 band shifted

to lower frequency relative to liquid benzene, the shift being
greater for larger ionic radius, which the authors interpreted as
being evidence for an interaction between the π-electrons of
benzene and the exchange cation. They also found (unlike pyridine) that the exact band position varied with benzene dosage and
concluded that the excess cations in X vs. Y zeolites resulted in
higher fields at the aromatic nucleus of benzene. These trends
were correlated with the selectivity of these catalysts for the
alkylation of toluene to give ethylbenzene and styrene.

In an earlier study Tam et al.(30,31,32) investigated the
adsorption of acetylene, dimethylacetylene and pyrazine (1,4-dia-
zine) on a series of alkali and alkaline earth exchanged X zeolites, and of acetylene on A type zeolites. A complete vibrational analysis of the five acetylene fundamentals was possible in
the case of NaA and CaA zeolites (30), demonstrating the versa-
tility of the Raman technique as a method of obtaining vibrational
fundamentals over a wide frequency range. For pyrazine (32), the
high frequency shift of the ring breathing mode relative to
liquid pyrazine was proportional to the polarizing power of the
cation [i.e., a smaller shift for increasing r] and independent
of the cation charge, a result which contrasts with the e/r
correlation found by Ward (33), and by us (8), for pyridine
adsorption [although qualitatively the r dependence is the same
for equal cation charge]. Tam and Cooney (32) noted that pyrazine is a much weaker base than pyridine since it has no dipole
moment and therefore would be less able to accommodate positive
charge than pyridine.

For acetylene adsorbed on exchanged A and X zeolites, Tam
et al. (30,31) found that the diagnostic C≡C stretching mode
[1974 cm-1 for gaseous C_2H_2] shifted downward upon adsorption,
the magnitude of the shift now being inversely proportional to
the cation polarizing power of the cation, i.e., a larger shift
for larger cation radius [as was found for Unland et al. (7) for
benzene]. A complex model was discussed in terms of adsorbate-
adsorbate interactions, cation-adsorbate attractions and oxide-
adsorbate repulsions.

It is worth noting that qualitatively the nitrogen heterocyclic systems (8,27,32) behave similarly in showing a high
frequency shift which decreases with increasing cation radius,
whereas benzene (7) and acetylene (30,31) show a low frequency
shift which increases with r. This probably arises because the
latter can only interact via a π-electron mechanism, whereas
interaction with the nitrogen lone pair electrons is probably
the dominant factor with pyridine and pyrazine. In a later study
of the adsorption of C_2H_2 on γ-alumina and KX zeolite, Heaviside
et al. (34) concluded that acetylene is physisorbed via its
π-electron system in a "side-on" orientation, perhaps to surface
OH groups or traces of retained water.

In a different application of Raman spectroscopy to studies
of zeolites, Cooney and Tsai (6) have investigated the adsorption
of bromine on alkali exchanged zeolites X, NaY and NaA, and the

subsequent reaction with benzene on NaX and CsX (35). On zeolite X, Br_2 adsorbed on the more exposed III sites (νBr_2=280-314 cm^{-1}) and on the less exposed II sites (νBr_2=240-280 cm^{-1}), the latter also having an $\nu(0...Br)$ band at $\sim$ 160 cm^{-1} indicative of an oxide-bromine electron transfer interaction. No site III cations exist for NaY and only the site II NaY-Br_2 band was observed ($\sim$ 271 cm^{-1} for NaX and NaY) while for NaA, both site II (281 cm^{-1}) and I (259 cm^{-1}) bands were observed, but in neither case was a low frequency band observed near 160 cm^{-1} indicating, in agreement with others, the apparent lack of charge transfer interactions.

The subsequent reaction of bromine treated NaX and CsX with benzene revealed two types of behavior (35). At saturation Br_2 coverage surface donor complexes were formed on sites III and II, whereas at less than Br_2 saturation (only site II occupied) benzene reacted rapidly to form addition products containing carbon-bromine bonds. The unique ability to use Raman spectroscopy in general for obtaining low frequency spectral data in studies of in situ catalytic process was discussed by the authors.

Finally, in a brief communication Saperstain and Rein (26) have reported that Raman spectroscopy can even be used to detect the presence of physically adsorbed N_2 and O_2 on a 4A zeolite, and the results are discussed in terms of the chromatographic ability of the A zeolites to separate N_2 from other gases. This ability to detect adsorbed gases on zeolite might prove useful for in situ catalytic studies, but thus far no additional uses of the technique have been reported.

Chemisorption on Oxides

The work described in the previous section was essentially concerned with the physical rather than chemical adsorption of some highly polarizable molecules on zeolites. With pyridine, such information can sometimes be obtained as easily using infrared spectroscopy. However, transmission IR spectroscopy cannot so easily be used to study chemisorption on oxides if it is essential to obtain low frequency spectral data (e.g., adsorbent-adsorbate stretching modes) because of the opacity of most oxides over much of the low frequency spectral region. Recent work has shown that the Raman technique can be extremely useful in this context (4).

Surface hydroxyl groups on an oxide can usually be replaced by other organic functional groups, thereby altering the polarity or hydrophobicity of the surface. One simple process involves the dissociative chemisorption of methanol on silica.

$$\equiv SiOH + CH_3OH \rightarrow \equiv SiOCH_3 + H_2O$$

The methoxylation can be carried out by reacting silica with methanol vapor at 300-400°C, or by refluxing silica in methanol (21,36). Because the infrared spectrum of the modified surface is well understood (36) we chose to use this system as a model to test the feasibility of using Raman spectroscopy (21) for studying such surface modification procedures.

Raman spectra obtained in the spectral region above 1200 cm^{-1} were of a quality which was comparable to that which could be obtained using infrared spectroscopy, although longer scan times and larger spectral slit widths were required in order to achieve a comparable signal-to-noise ratio (Figure 4). However, we were unable to observe low frequency Raman bands (Figure 2), particularly those associated with the ν(Si-O-C) modes, a most disappointing result because it is in this low frequency region (< 1200 cm^{-1}) where the Raman technique should prove so advantageous over the infrared. A possible cause for our failure to detect these modes will be discussed further below.

Surface hydroxyl groups also react with a class of molecules sometimes referred to as hydrogen sequestering (HS) agents, examples of which are $TiCl_4$, $AlMe_3$ [Me = CH_3], $Me_3SiNHSiMe_3$ [HMDS or hexamethyldisilazane], $MCl_{4-n}Me_n$ [n = 0-3, M = Si or Ge] and BX_3 [X = F, Cl, Br]. Not only are some of these molecules used to modify the surface properties of an adsorbent or to impart new reactive sites on a catalyst, they have also been widely used as probe molecules to study the configurations of surface hydroxyl groups (37 - 43). For example, $TiCl_4$ might be expected to react differently with single surface OH groups or with "paired" OH groups as follows (M represents a metal atom of an oxide):

single

$$MOH + TiCl_4(g) \rightarrow MOTiCl_3 + HCl(g) \qquad\qquad I$$

pairs (H bonded)

$$\begin{array}{l} MOH \\ \\ MOH \end{array} + TiCl_4(g) \rightarrow \begin{array}{l} MO \\ \diagdown \\ MO\diagup \end{array}TiCl_2 + 2HCl(g) \qquad II$$

(geminal) $\quad M(OH)_2 + TiCl_4(g) \rightarrow MO_2TiCl_2 + 2HCl(g) \qquad III$

These reactions have been extensively studied using classical analytical techniques and using IR spectroscopy (37-43). In the latter case, because of the low frequency opacity of the oxide one can generally only observe spectra associated with the disappearance of ν(OH) modes during reaction, or the formation of CH stretching modes if a methyl group is present, and in some cases it is difficult to distinguish between reactions illustrated in schemes II and III.

We have used Raman spectroscopy (4) in order to see if low frequency data could be obtained when the above HS agents dissociatively chemisorb on the isolated hydroxyl groups of silica (scheme I). As was found using methanol, very good quality Raman spectra in the CH stretching and deformation regions could be obtained when a methyl group was present in the HS agent, and over the entire spectral region ($\sim$ 50-4000 cm^{-1}) for the adsorbed germanium compounds, HMDS, $AlMe_3$, and $TiCl_4$. The spectra were analyzed in terms of the surface product predicted in scheme I. For example, Figure 5 shows the low frequency portion of the Raman spectrum of adsorbed $GeClMe_3$(A), $GeCl_2Me_2$(B) and $GeCl_3Me$(C) where the chemisorbed species are respectively $\equiv SiOGeMe_3$(A), $\equiv SiOGeClMe_2$(B) and $\equiv SiOGeCl_2Me$(C). The bands between 600-700 cm^{-1}

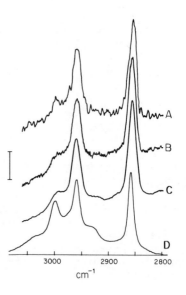

Figure 4. Raman spectra of methylated silica in the CH stretching region using 1.7 W of laser power at 488.0 nm. The scan rate and response time constant (seconds) were as follows: (A) 500 cm⁻¹/ min, 0.1 s; (B) 50 cm⁻¹/min, 1 s; (C) 10 cm⁻¹/min, 10 s. Curve D shows the IR spectrum recorded with a scan time of 83 cm⁻¹/min and a time constant of 1 s. The vertical bar represents 3000 Hz for the Raman spectra and a transmittance of 0.20 for the IR spectrum (21).

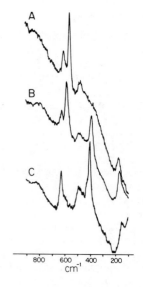

Figure 5. Raman spectra of chemisorbed (A) GeClMe₃, (B) GeCl₂Me₂, and (C) GeCl₃Me. Using 0.7 W of laser power (488 nm), the maximum intensity is as follows: (A) 4600 Hz at 573 cm⁻¹; (B) 1000 Hz at 595 cm⁻¹; (C) 3000 Hz at 410 cm⁻¹ (4).

are due to GeC stretching modes, those between 400-450 cm^{-1} due to GeCl stretching modes, the broad band between 100-200 cm^{-1} is associated with an overlapping system of complex GeC and GeCl deformation modes, and the accentuated peak near 480-490 cm^{-1} (which is also observed weakly in the background spectrum of silica) has been assigned to the symmetric SiOGe stretching mode. Other spectra and a complete discussion of the assignments are given in reference (4).

Very surprisingly, no Raman bands at all were observed for the adsorbed boron compounds, in spite of the fact that the band position for some of these were known from IR studies (40,43). This was also the case for the low frequency modes of chemisorbed methanol (21). We have speculated that this effect may arise when the mass of the atom M attached to the surface oxygen forming the ≡SiOM linkage is less than that of silicon, such that the low frequency modes are strongly coupled to the lattice or phonon modes of the substrate resulting in a broad smeared-out profile, which is essentially undetectable in the general background at low frequency.

As is well known, vibrational modes which are intense in the Raman effect are often weak in the infrared, and *vice versa*. This was true in the above study, where the antisymmetric MC and MCl bands were almost undetectable in some cases, and in order to carry out a complete vibrational analysis it would also be desirable to obtain the infrared spectrum. It has been pointed out that this is difficult using transmission techniques because oxides are such strong IR absorbers at low frequency. However, the extraction of weak signals which are superimposed on an intensely absorbing background can sometimes be realized using a highly sensitive dispersive or Fourier transform spectrometer.

Figure 6A shows a portion of the Raman spectrum of chemisorbed hexamethyldisilazane on silica, a reaction in which the surface hydroxyl groups are replaced by ≡SiOSiMe$_3$ groups. The strong and weak features at 600 and 690 cm^{-1} are due to the symmetric and anti-symmetric ν(SiC) modes respectively whereas the weaker features at 755 and 850 cm^{-1} are due to methyl rocking modes. Figure 6B shows the corresponding infrared spectrum. Silica is totally opaque from 1200-1000 cm^{-1} and the transmittance is only 2.5% from 850-800 cm^{-1}, but within the "window" of partial transparency to high frequency of the 850-800 region one can clearly discern an extra band due to a methyl twisting mode which is absent in the Raman spectrum. The methyl rocking and twisting modes are expected to be weak in the Raman effect so the combination of the two spectroscopies should, in favorable cases, go a long way towards enabling one to carry out a full vibrational analysis. Although this type of application of IR spectroscopy is a new development, and the technique is tedious since extremely slow scan times are required, we have recently been able to observe similar modes for the chemisorbed Ge compounds. In some cases we have also observed the pseudoantisymmetric SiOGe stretching mode, a mode which was not observed in the Raman

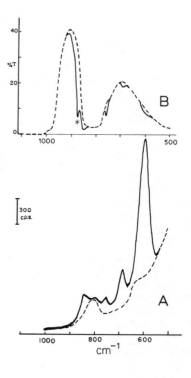

Figure 6. (A) Raman spectrum and (B) IR spectrum of chemisorbed hexamethyldisilazane on silica: (– – –) the background spectrum in each case; (———) new features after formation of surface ≡SiOCH₃ groups. These are downward relative to background in the IR spectrum B and upward in the Raman Spectrum A, and the horizontal line for Curve B indicates total absorption for the IR spectrum. The band marked (*) does not appear in the Raman spectrum.

spectrum, probably because the change in polarizability is so low. The infrared spectrum shown in Figure 6B serves to show why the Raman technique is so useful in the low frequency spectral regions.

However, in the higher frequency spectral regions where most oxides are relatively transparent to IR radiation, the combined infrared and Raman technique has been very successfully used by others to study chemisorption on zinc oxide (44,45,46). One also makes use of the fact that the selection rules differ for each technique, and hence the relative intensities vary. Nguyen and Sheppard (44) found that propylene gave a chemisorbed allyl species on this oxide whereas Nguyen (45) found that ethynyl-benzene (phenylacetylene) dissociatively chemisorbed to yield an acetylide species, probably Zn...C≡CC$_6$H$_5$, accompanied by dimeric products which were generated only after adsorption and laser irradiation. In a more detailed work Nguyen, Lavalley, Saussey and Sheppard (46) investigated the chemisorption of methylphenyl-acetylene (C$_6$H$_5$C≡CCH$_3$) and benzylacetylene (C$_6$H$_5$CH$_2$C≡CH) on zinc oxide. With either adsorbate, two types of dissociative chemi-sorption occurred, giving rise to acetylide species (C$_6$H$_5$CH$_2$C≡C...Zn) and the propargyllic species [C$_6$H$_5$CH⋯CH⋯CH]⁻ and [C$_6$H$_5$C⋯C⋯CH$_2$]⁻, and the results were discussed in terms of an isomerization mechanism involving the formation of the other isomer, phenylallene (C$_6$H$_5$CH=C=CH$_2$).

Curiously, however, spectra due to some of the intermediates were not observed using both spectroscopic techniques, and the authors discuss possible causes for this. As with the previously discussed acetylene-zeolite systems (30,31), very strong Raman bands were observed for the ν(C≡C) modes which are very weak in the infrared (and forbidden in C$_2$H$_2$).

Finally, in another interesting application of the Raman technique, Murray and Graytak (5) recently observed the spectrum of ammonia adsorbed on silica (porous glass) under UHV conditions at coverages as low as 0.01 monolayer. Spectra were obtained of H-bonded NH$_3$, coordinated NH$_3$, and of surface SiNH$_2$ groups produced from the dissociative chemisorption of ammonia. In addition, they studied the rate of exchange of NH$_3$ and ND$_3$ with surface SiOD and SiOH groups respectively, and were able for the first time to use the Raman technique to monitor changes in the SiO stretching modes of ≡SiOH groups (∿ 980 cm⁻¹) as a function of coverage with NH$_3$. Although this study was more concerned with physical than chemical adsorption, it did show that extremely good quality Raman spectra can be obtained for a simple molecule under very carefully controlled UHV experimental conditions. This potential can possibly give the Raman technique the same status for studying oxides as the UHV electron techniques have for studying metals.

The Surface Structure of Oxide Catalysts

The work described in the previous two sections has been concerned with obtaining the Raman spectrum of molecules adsorbed

on oxide surfaces without much consideration of how the adsorbent
modes may themselves be perturbed. These investigations were
feasible because aluminosilicate oxides are relatively poor
Raman scatterers and it has been a simple matter to observe the
Raman spectrum of an adsorbed molecule. However, the low
intensity of the background scattering has proven to be useful in
recent Raman studies of silica and alumina supported molybdenum
oxides (9,10,19,20,47,48) since the scattering cross section of
the latter is high. Accordingly, several groups have used the
Raman technique to probe the nature of these catalyst surfaces
themselves, work which is important since these materials, particularly when promoted by adding oxides of cobalt or nickel, are
widely used as hydrosulfurization (HDS) catalysts.

It is not possible to review in detail the Raman work which
has been undertaken on these catalyst systems because a full
account would require a detailed description of the catalytic
properties of these substances, the chemistry of the molybdenum
oxides in general, and a full analysis of vibrational modes of
the $(Mo_xO_y)^{-n}$ anions. Briefly one can say that exceptionally good
quality but complex Raman spectra, an example of which is shown
in Figure 7, have been obtained in the frequency range from 50-
1100 cm^{-1}. Acknowledging that the "group frequency" concept may
be difficult to apply to the molybdenum oxides, Jeziorowski and
Knözinger (10) have concluded that Raman bands in this region may
be assigned as follows: 310-370 cm^{-1} (Mo=O bend), 900-1000 cm^{-1}
(Mo=O stretch), 200-250 cm^{-1} (Mo-O-Mo deformation), 400-600 cm^{-1}
(symmetric Mo-O-Mo stretch) and 700-850 cm^{-1} (anti-symmetric
Mo-O-Mo stretch). On this basis several groups (9,10,19,20,47,48)
have carried out detailed Raman studies varying such parameters as
(i) the method of catalyst preparation (pH during wet impregnation,
dry impregnation, calcining conditions), (ii) the catalyst support,
(iii) the ratio of molybdena to support, and (iv) the effect of
promoters, the objective being to understand how these variables
influence the structure of the molybdena surface, sometimes under
catalytic HDS conditions. Particular attention has been devoted
to obtaining an understanding of the conditions which result in
the formation of such species as solid MoO_3, polymeric oxyanions,
surface interactions (e.g., the formation of $Al_2(MoO_4)_3$ and of
surface Al-O-Mo bonds), mixed Mo-Co-Oxide species when a promoter
is used, and sulfide species (e.g., MoS_2) under desulfurization
conditions.

The spectra obtained, although complex, have provided a
valuable insight into the catalytic chemistry of these materials,
information which is particularly difficult to obtain using other
techniques. More important, the Raman technique can also be used
under real reaction conditions, and it has been used to examine
actual "reactor" catalyst samples. It is also apparent that these
techniques can be used to study other types of catalysts. For
example, Kerkhoff et al. (49) have obtained the Raman spectrum of
an alumina-supported rhenium oxide metathesis catalyst prepared
by calcining $(NH_4)ReO_4$ on γ-Al_2O_3 at 830 K. An extremely simple

spectrum was obtained (50-1200 cm^{-1}) which revealed that a single rhenium species was present, probably a tetrahedrally distorted ReO$_4^-$ ion. As yet no attempt has been made to study this surface under reaction conditions but it is apparent that these techniques should receive much further application in the future.

Finally, we have not discussed cases where Raman spectroscopy can be used to study catalysts indirectly, as for example, by extracting a sample from a reactor and preparing a KBr disc for IR or Raman investigation. Such techniques may be useful in special circumstances (50) but have limited applicability with regard to the direct examination of surfaces under reaction conditions.

Adsorption on Metals

Ten years ago one would have predicted that Raman spectroscopy could never be used to study monolayer adsorption on metals because either (a) the sensitivity of the technique would be too low to permit detection of signals using a single reflection from a smooth metal surface, or (b) oxide supported metal surfaces are black if the metal loading is high and therefore the laser light would be absorbed. Both of the above objections have been shown to be faulty insofar as the technique has now been used to study absorption on silica-supported nickel (51,52,53) and on single crystal nickel (54). Moreover in the special case of silver, Raman spectra have been observed wherein the signals obtained are of the order of 10^4-10^6 greater than one would predict on the basis of the normal Raman scattering cross section of bulk molecules (11,12). The latter effect has sometimes been termed "enhanced Raman scattering" and has been the subject of intensive theoretical and experimental investigation. We will first discuss the normal or general situation.

(i) Normal Raman Scattering. Because of the objections listed above it has proven difficult to use the Raman technique to study metals. Krasser et al. have had some success in obtaining spectra of carbon monoxide adsorbed on silica-supported Ni and on Raney Ni (51), and of hydrogen (52,53) and benzene (53) on silica-supported Ni. The spectra are very weak for adsorbed CO (signal-to-noise ratio $\sim$ 4 for the strongest band, but the absolute intensity was not stated) and the peak positions and intensities varied with CO pressure, the state of focus of the laser beam, and on the length of exposure of the sample to the laser. In addition to ν(C$\equiv$O) bands assigned as usual to linear (> 2000 cm^{-1}) and bridged (< 2000 cm^{-1}) CO (3), a weak low frequency band at 340 cm^{-1} was assigned to an ν(NiC) mode, and the authors claimed that adsorbed Ni(CO)$_4$ was also present. There was little discussion of how some of the above variables affected the spectrum and it would appear that the technique has not been fully exploited.

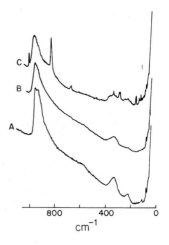

*Figure 7.   Raman spectrum of an 8 wt %
molybdena on γ-Al₂O₃ (A) after impregnation at pH 6, (B) after drying at 393 K,
(C) after calcination at 773 K. Redrawn
from Ref. 10.*

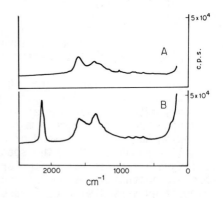

*Figure 8.   Raman spectra of: (a) a
freshly polished Ag sample exposed to
air, (b) a Ag sample in air after exposure
for 30 min to a 3.7% KCN solution. Redrawn from Ref. 63.*

For adsorbed H_2 (and D_2) Krasser and Renouprez (52) found several extremely weak Raman bands in the range 600-2100 cm^{-1} which were assigned to linear NiH species (2028 and 1999 cm^{-1}) and multibonded Ni_xH species at 1600 and 950 cm^{-1} and again no studies of the spectral changes with respect to reactivity of the NiH species with other molecules were carried out.

Stencel and Bradley (54) carried out an elegant study of the adsorption of CO on Ni(100, 110 and 111) surfaces under UHV conditions although signal intensities were only of the order of 2-20 c.p.s. (compared with $\sim 1 \times 10^4$ for chemisorption on high area oxides). Spectra were obtained only in the region associated with $\nu(C\equiv O)$ modes and although the authors acknowledged that their crystals may not have been totally free of oxide, no bands normally associated with surface carbonates (950-1700 cm^{-1}) were detected. Further, no Raman bands were observed in the 1940-2000 cm^{-1} region associated with the formation of bridged CO species. Different $\nu(NiCO)$ spectra (2000-2050 cm^{-1}) were obtained for each crystal plane but the data was not discussed in detail in this preliminary communication because of doubts associated with the state of cleanliness of the surface.

Finally, two papers have been concerned with a study of adsorption on platinum electrodes in aqueous solution. One was concerned with the adsorption of CO (55) and the other with I_2 (56), and although again very weak signals were detected [$\nu(PtCO)$ at 2096 and 2081 cm^{-1}, and $\nu(I-I)$ at 174 cm^{-1}], the experiments do illustrate one of the major advantages of the Raman technique, viz. the ability to study adsorption from aqueous media.

(ii) Enhanced Raman Scattering. Raman spectra which were about 10^4 - 10^6 more intense than anticipated were first observed (57-62) for pyridine adsorbed on silver electrodes in an electrochemical situation and subsequently for other heterocyclic molecules (60) and ions such as CN$^-$ (63,64,65,66) $CO_3^=$, (60) Cl$^-$ (11) and SCN$^-$ (67). It was originally thought that this effect was an artefact due to the applied field generated in the electrochemical deposition but similar enhanced spectra have been observed for CO (18) and pyridine (17) deposited on silver under UHV conditions. Although Raman enhancement has now been demonstrated many times for adsorption on silver, only recently has it also been shown to occur for electrochemically deposited pyridine on copper and gold electrodes (68,69). To this author's knowledge Raman enhancement has not been achieved using other metals. The spectrum shown in Figure 8 for adsorbed CN serves to illustrate the quality of the spectra obtained with respect to peak intensity and signal to noise ratio. In studies using pyridine, spectra of the quality shown in Figure 3A for liquid pyridine have been observed.

It is too early to say whether surface enhanced Raman spectroscopy will become a widely applicable technique because, in spite of several theoretical investigations, no general theory of

the effect has been forthcoming which would point to the conditions necessary in order to study adsorption on other metals.
Recent reviews of theories of the effect have been given by Hexter
(11) and by Efrima and Metiu (12) and, since several new papers
are currently appearing each month, it is not appropriate in a
short article that we discuss this work in detail. Briefly, most
theories assume that a resonance Raman scattering effect is
responsible for the enhancement and models have been developed in
terms of (i) classical (12,64,70-73) and quantum mechanical (74)
descriptions of light scattering as related to the optical
properties of metals, (ii) the role of surface roughness in promoting a resonant excitation of conduction electron resonances
in an adsorbate (75), or in promoting a breakdown in certain
momentum conservation rules (65), (iii) a mixing of surface
plasmons with adsorbate states (11,68,76,77) and (iv) resonance
excitation of free radical species at an electrode surface (78).
        These theoretical developments are hampered by conflicting
experimental evidence with respect to the conditions necessary to
generate the effect. For example, there is no clear consensus on
the effect of changing the wavelength of the exciting radiation
(11,12), and some groups claim that electrochemical "roughening"
is a necessary precursor to observing the effect (63,65,66,68,79,
80,81). However, the effect has clearly been demonstrated under
UHV conditions for CO adsorbed on evaporated silver films (18) and
for pyridine adsorbed on ion sputtered bulk silver (17) and it
remains to be seen whether the necessary surface "roughening"
condition may have been met. Finally, an adequate theory must
account for the following experimental observations; (i) only a
limited number of adsorbed molecules on silver give rise to the
enhancement effect, (ii) enhancement for pyridine in solution
apparently does not occur with non-aqueous solvents (61),
(iii) using silver electrodes the signal intensity is strongly
dependent on the anodization time (63,66,68,69,79,80,81) and
(iv) for much less than a monolayer of pyridine on an Ag electrode
the intensity of the scattered light (for a constant 90° illumination-scattering system) strongly depends on the angle of incidence
(63°, half width 5° for a silver film on glass) which Pettinger
et al. (68,77) claim provides strong evidence for the surface
plasmon enhanced model for the effect. Very recently a similar
angular dependence on the scattering intensity has also been
found by Wenning, Pettinger and Wetzel (69) for pyridine adsorbed
on copper and gold electrodes. Most interestingly however, Raman
enhancement for these metals was only achieved using red (647 nm)
laser excitation whereas for silver, enhancement occured with red
or blue (457 nm) light.
        The resolution of the problems outlined above will obviously
require much more theoretical and experimental work before the
Raman technique will become universally applicable to the study of
adsorption at metal surfaces. It will be interesting to see if
tuneable lasers capable of operating in the near infrared will be

the key to extending the use of this technique to other metals.
Indeed, if spectra of the quality shown in Figure 8 could be
obtained for any metal, other vibrational techniques, at least as
applied to the study of adsorption on metals, could be rendered
obsolete. Equally important, the technique can be used to study
adsorption from aqueous solution where it is virtually impossible
to obtain vibrational data for adsorbed species using any other
technique.

Summary

We have demonstrated that Raman spectroscopy is a versatile
technique for studying physical and chemical adsorption on oxides,
the surface structure of certain types of supported oxide
catalysts, and in special circumstances, for studying adsorption
on metals. Of all vibrational techniques for surface studies it
is unique in that (i) there are virtually no restrictions on the
type of sample that can be studied (ii) for gas-solid studies
there are no restrictions on the conditions of ambient gas
pressure, (iii) the full range of vibrational fundamentals
($\sim$ 50-4000 cm^{-1}) is accessible under relatively high resolution
conditions and (iv) adsorption can be studied from aqueous solutions on to both oxide and metal surfaces. Although the sensitivity of the technique is not always as high as one would like,
[with some exceptions (5), and particularly the enhanced effect
for silver] this can be partially overcome by the use of a data
acquisition computer which permits signal averaging, and with the
use of several lasers of a tuneable laser in order to minimize
background fluorescence.

In the future, the technique will undoubtedly be further
exploited along the lines described, particularly for the study of
in situ catalytic processes and, if the experimental and theoretical problems can be solved, for the study of adsorption on well
defined metals. As yet the technique has not been fully used for
study of adsorption from aqueous solution and, for example, there
is considerable potential for studying the role of sediments in
transporting pollutants.

Raman spectroscopy has also shown considerable promise as a
means of studying the generation of defects in zeolites as a
result of various thermal or chemical treatments (82). Cooney and
Tsai (6) have correctly pointed out that although X-ray diffraction is useful for studying ordered systems, it is not generally
capable of detecting disordered domains. On the other hand, the
Raman technique is capable of studying both ordered and disordered
domains and since highly crystalline zeolites often exhibit sharp
characteristic bands in the low frequency region, small changes
due to thermally induced surface defects are readily detected, as
has been found in the infrared spectrum of amorphous silica (83).
Murray and Greytak (5) have also discussed similar situations in
the case of the Raman spectrum of porous glass. The ability to

use Raman spectroscopy to study zeolites under various reaction conditions should prove useful in future studies of these effects.

Acknowledgement

Financial support for part of this work was provided by N.S.E.R.C. of Canada and Imperial Oil Limited in collaboration with Dr. A.H. Hardin, Dr. W.N. Sont, Mr. A. St. Onge and Mrs. M. Klemes.

Literature Cited

1. Kiselev, A.V., Lygin, V.I. "Infrared Spectra of Surface Compounds", Wiley, New York, 1975.
2. Hair, M.L. "Infrared Spectroscopy in Surface Chemistry", Marcel Dekker, New York, 1967.
3. Sheppard, N., Nguyen, T.T. Adv. Infrared Raman Spectrosc., 1978, 5, 67.
4. Morrow, B.A., Hardin, A.H. J. Phys. Chem., 1979, 83, 3135.
5. Murray, C.A., Greytak, T.J. J. Chem. Phys., 1979, 71, 3355.
6. Cooney, R.P., Tsai, P.J. Raman Spectrosc., 1979, 8, 195.
7. Freeman, J.J., Unland, M.L. J. Catal., 1978, 54, 183.
8. Hardin, A.H., Klemes, M., Morrow, B.A. J. Catal., 1980, 62, 316.
9. Cheng, C.P., Schrader, G.L. J. Catal., 1979, 60, 276.
10. Jeziorowski, H., Knözinger, H. J. Phys. Chem., 1979, 83, 1166.
11. Hexter, R.M. Solid State Comm., 1979, 32, 55.
12. Efrima, S., Metiu, H. Israel J. Chem., 1979, 18, 17.
13. Egerton, T.A., Hardin, A.H. Catal. Rev. Sci.-Eng., 1975, 11, 1.
14. Cooney, R.P., Curthoys, G. Nguyen, T.T. Adv. Catal., 1975, 24, 293.
15. Takenaka, T. Adv. Coll. Interface Sci., 1979, 11, 291.
16. Delgass, W.N., Haller, G.L., Kellerman, R., Lunsford, J.H. "Spectroscopy in Heterogeneous Catalysis", Academic Press Inc. New York, 1979, p. 58-85.
17. Smardzewski, R.R., Colton, R.J., Murday, J.S. Chem. Phys. Lett., 1979, 68, 53.
18. Wood, T.H., Klein, M.V. J. Vac. Sci. Technol. 1979, 16, 459.
19. Brown, F.R., Makovsky, L.E., Rhee, K.H. J. Catal., 1977, 50, 162.
20. Medema, J., van Stam, C., de Beer, V.H.J., Konigs, A.J.A., Koningsberger, D.C. J. Catal., 1978, 53, 386.
21. Morrow, B.A., J. Phys. Chem., 1977, 81, 2663.
22. Pettinger, B., Wenning, U., Wetzel, H. Chem. Phys. Lett., 1979, 67, 192.
23. Greenler, R.G., Slager, T.L. Spectrochim. Acta, 1973, 29A, 193.
24. Egerton, T.A., Hardin, A.H., Kozirovski, Y., Sheppard, N. J. Catal., 1974, 32, 586.
25. Jeziorowski, H., Knözinger, H. Chem. Phys. Lett., 1977, 51, 519.

26. Saperstein, D.D., Rein, A.J. J. Phys. Chem., 1977, 81, 2134.
27. Egerton, T.A., Hardin, A.H., Sheppard, N. Can.J.Chem.,
    1976, 54, 586.
28. Ward, J.W. J. Catal., 1968, 10, 34.
29. Ward, J.W. J. Colloid Sci., 1968, 28, 269.
30. Tam, N.T., Cooney, R.P., Curthoys, G. J. Chem. Soc.,
    Faraday I, 1976, 72, 2577.
31. Tam, N.T., Cooney, R.P., Curthoys, G., J. Chem. Soc.,
    Faraday I, 1976, 72, 2592.
32. Tam, N.T., Cooney, G. J. Chem. Soc., Faraday I, 1976, 72,
    2598.
33. Ward, J.W. J. Catal., 1969, 14, 365.
34. Heaviside, J., Hendra, P.J., Tsai, P., Cooney, R.P.
    J. Chem. Soc. Faraday I, 1978, 74, 2542.
35. Tsai, P., Cooney, R.P. J. Raman Spectrosc., 1979, 8, 236.
36. Morrow, B.A. J. Chem. Soc., Faraday I, 1974, 70, 1527.
37. Hair, M.L. J. Colloid Interface Sci., 1977, 60, 154.
38. Hair, M.L., Hertl, W. J. Phys. Chem., 1973, 77, 2070.
39. Armistead, C.G., Tyler, A.J., Hambleton, F.H., Mitchell, S.A.,
    Hockey, J.A. J. Phys. Chem., 1969, 73, 3947.
40. Morrow, B.A., Devi, A. J. Chem. Soc., Faraday I, 1972,
    68, 403.
41. Evans, B., White, T.E. J. Catal., 1968, 11, 336.
42. Tyler, A.J., Hambleton, F.H., Hockey, J.A., J. Catal.,
    1969, 13, 35.
43. Bermudez, V.M. J. Phys. Chem., 1971, 75, 3249.
44. Nguyen, T.T., Sheppard, N. J. Chem. Soc., Chem. Comm.,
    1978, 868,
45. Nguyen, T.T. J. Catal., 1980, 61, 515.
46. Nguyen, T.T., Lavalley, J.C., Saussey, J., Sheppard, N.
    J. Catal., 1980, 61, 503.
47. Knözinger, H., Jeziorowski, H. J. Phys. Chem., 1978, 82,
    2002.
48. Brown, F.R., Makovsky, L.E., Rhee, K.H. J. Catal., 1977,
    50, 385.
49. Kerkhof, F.P.J.M., Moulijn, J.A., Thomas, R. J. Catal.,
    1979, 56, 279.
50. Hoefs, E.V., Monnier, J.R., Keulks, G.W. J. Catal., 1979,
    57, 331.
51. Krasser, W., Ranade, A., Koglin, E., J. Raman Spectrosc.,
    1977, 6, 209.
52. Krasser, W., Renouprez, A.J. Raman Spectosc., 1979, 8, 92.
53. Krasser, W., Renouprez, A. Proc. Int. Conf. on Vibrations
    in Adsorbed Layers, Jülich, D.B.R. (June 1978), p. 175.
54. Stencel, J.M., Bradley, E.B. J. Raman Spectrosc., 1979,
    8, 203.
55. Cooney, R.P., Fleischmann, M., Hendra, P.J., J. Chem. Soc.
    Chem. Comm., 1977, 235.
56. Cooney, R.P., Reid, E.S., Hendra, P.J., Fleischmann, M.
    J. Chem. Soc., Chem. Comm., 1977, 2002.

57. Fleischmann, P., Hendra, P.J., McQuillan, A.J. Chem. Phys. Lett., 1974, 26, 163.
58. McQuillan, A.J., Hendra, P.J., Fleischmann, M. J. Electroanal. Chem., 1975, 65, 933.
59. Fleischmann, M., Hendra, P.J., McQuillan, A.J., Paul, R.L., Reid, E.S. J. Raman Spectrosc. 1976, 4, 269.
60. Jeanmarie, D.L., van Duyne, R.P., J. Electroanal. Chem., 1977, 84, 1.
61. van Duyne, R.P. J. Phys. (Paris), 1977, C5, 239.
62. Albrecht, M.G., Creighton, J.A. J. Am. Chem. Soc., 1977, 99, 5215.
63. Otto, A. Surface Sci., 1978, 75, L392.
64. Otto, A. Surface Sci., 1980, 92, 145.
65. Billmann, J., Kovacs, G., Otto, A. Surface Sci., 1980, 92, 153.
66. Bergman, J.G., Heritage, J.P., Pinczuk, A., Worlock, J.M., McFee, J.H. Chem. Phys. Lett., 1979, 68, 412.
67. Cooney, R.P., Reid, E.S., Fleischmann, M., Hendra, P.J. J. Chem. Soc., Faraday I, 1977, 73, 1691.
68. Pettinger, B., Wenning, U., Wetzel, H. Chem. Phys. Lett., 1979, 67, 192.
69. Wenning, U., Pettinger, B., Wetzel, H. Chem. Phys. Lett., 1980, 70, 49.
70. Efrima, S., Metiu, H. J. Chem. Phys., 1979, 70, 1602.
71. Efrima, S., Metiu, H. Surface Sci., 1980, 92, 417.
72. King, F.W., van Duyne, R.P., Schatz, G.C. J. Chem. Phys., 1978, 69, 4472.
73. Efrima, S., Metiu, H. J. Chem. Phys., 1979, 70, 1939.
74. King, F.W., Schatz, G.C. Chem. Phys., 1979, 38, 245.
75. Moskovits, M., J. Chem. Phys., 1978, 69, 4159.
76. Hexter, R.M., Albrecht, M.G. Spectrochim. Acta, 1979, 35A, 233.
77. Pettinger, B., Tadjeddine, A., Kolb, D.M. Chem. Phys. Lett., 1979, 66, 544.
78. Regis, A., Corset, J. Chem. Phys. Lett., 1980, 70, 305.
79. Albrecht, M.G., Evans, J.F., Creighton, J.A. Surface Sci. 1978, 75, L777.
80. Pettinger, B., Wenning, U. Chem. Phys. Lett., 1978, 56, 253.
81. Kirtley, J.R., Tsang, J.C. Bull. Am. Phys. Soc., 1979, 24, 278.
82. Morrow, B.A., Sont, W.N. unpublished results.
83. Morrow, B.A., Lee, L.S.M., Cody, I.A. J. Phys. Chem., 1976, 80, 2761.

RECEIVED June 3, 1980.

8

Investigations of Adsorption Centers, Molecules, Surface Complexes, and Interactions Among Catalyst Components by Diffuse Reflectance Spectroscopy

K. KLIER

Lehigh University, Sinclair Laboratory, Bethlehem, PA 18015

Diffuse reflectance spectroscopy (DRS) is a method for obtaining light intensity loss spectra in absorbing and scattering specimens and, with the use of appropriate theories, for extracting information on the molar absorptivities of species involved in the light absorption process. Extensive monograph (1-5) and review (6-8) literature is available to cover both the technical details and applications. The various configurations of spectrometers for DRS have been thoroughly discussed in Kortüm's book (1), and cells for air-sensitive samples such as catalysts have been reviewed in detail in references 5 and 8. The wavelengths of the radiation probe range from infrared to ultraviolet and both vibrational and electronic spectral measurements of powdered solids can be readily accomplished in combination with standard techniques of sample preparation and protection. Among the recent instrumental advances the most significant appears to be computerization of reflectance spectrometers which allows, in addition to an easy quantitative data processing, small quantities of adsorbed species to be detected and reflectance measurements to be made on very dark samples. We will not review in detail these instrumental arrangements, which consist at the present stage of individual interfacing of various spectrophotometers with micro-, mini-, and large computers and certainly have not reached a mature commercial stage of development. Rather, after an exposure of the basic model and radiative transfer theory of the absorbing-scattering media, we shall proceed to those applications which have successfully contributed to the resolution of structures, dynamics, and electronic states of adsorbates and catalysts.

Theory

A comprehensive theory of radiative transfer, which governs the radiation field in a medium that absorbs, emits, and scatters

0-8412-0585-X/80/47-137-141$05.50/0
© 1980 American Chemical Society

radiation, has been formulated and elaborated by S. Chandrasekhar
(9). The fundamental Chandrasekhar's equation of radiative transfer, which is applicable to scattering-absorbing-emitting media
of all possible geometries and spatial distributions of absorbers,
scatterers, and emitters, reads as follows:

$$-\frac{dI_\nu}{\kappa_\nu \rho ds} = I_\nu - (j_\nu/\kappa_\nu). \tag{1}$$

Here I_ν is the irradiance (or intensity) at a point (x,y,z) and in
a direction s of frequency between ν and $\nu + d\nu$, ρ is the local
density of the medium and ds is elemental pathlength in the direction s. The mass scattering-mass absorption coefficient κ_ν
describes the relative loss of intensity upon passage of light
through ds due to scattering away from the direction s and attenuation of light by absorption along s; the emission coefficient j_ν
describes the gain of intensity in the direction s due to scattering from all other directions and emission from internal sources.
The angular distribution of the scattered radiation is specified
by the so called *phase function* $p(\cos \Theta)$ which is proportional to
the rate at which light is being scattered from the direction s
into s' such that $s \cdot s' = \cos \Theta$. Isotropic scattering has a phase
function $p(\cos \Theta) = \bar{\omega}_0$, where $\bar{\omega}_0$ is an angle-independent *albedo*
for single scattering and assumes values between 0 and 1. In
plane parallel isotropically scattering media without internal
sources, in which no unscattered light propagates, equation (1)
assumes the simple form

$$-\mu \frac{dI_\nu(z, \mu)}{\kappa_\nu \rho \, dz} = I_\nu(z, \mu) - \frac{1}{2} \bar{\omega}_0 \int_{-1}^{1} I_\nu(z, \mu')d\mu' \tag{2}$$

where μ is the cosine of the angle of propagation with respect to
the inward normal to the illuminated surface and z is the axis of
that normal with positive direction inward. Solutions of the
integro-differential equations (1) and (2) have been found by
Chandrasekhar for a variety of phase functions and boundary conditions. For each physical problem specified by the boundary
conditions the angular distribution of the light intensity
becomes known at any point of the medium.

    In diffuse reflectance spectroscopy the total fluxes, rather
than angular distributions of intensities, are measured. The
forward flux $F_\nu^+(z) = \int_0^1 \mu I_\nu(z, \mu) \, d\mu$ and the backward flux
$F_\nu^-(z) = -\int_1^0 \mu I_\nu(z, \mu) \, d\mu$ define the reflectance R from the front
(illuminated) surface of a plane parallel specimen as

$$R = F_\nu^-(0)/F_\nu^+(0)$$

and the transmittance T at a thickness Z as

$$T = F_\nu^+(Z)/F_\nu^+(0)$$

where $z = 0$ for the illuminated surface and $z = Z$ for the non-illuminated surface of the absorbing and scattering plane parallel medium. Equation (2) has been solved by Chandrasekhar for the case of semi-infinite plane parallel medium and the expressions for R and T have been obtained by Klier (10) in the form

$$R = \frac{1 + R_g(b \coth Y - a)}{a + b \coth Y - R_g} \quad (3)$$

$$T = \frac{b}{b \cosh Y + a \sinh Y} \quad (4)$$

Here $Y = \xi \kappa_\nu \rho Z$, where ξ is the positive real root of the characteristic equation

$$\bar{\omega}_o = \frac{2\xi}{\ell n[(1+\xi)/(1-\xi)]} \ ,$$

$a = -\dfrac{(1 + \phi^2)}{2\phi}$ and $b = \sqrt{a^2-1}$, where $\phi = \dfrac{\xi+\ell n(1-\xi)}{\xi-\ell n(1+\xi)}$,

and R_g is the reflectance of the background placed at the non-illuminated surface

$$R_g = F_\nu^-(Z)/F_\nu^+(Z).$$

Equations (3) and (4) are formally identical with the earlier Kubelka's hyperbolic solutions of differential equations for forward and backward fluxes (11), although the Chandrasekhar-Klier and Kubelka's theories start from different sets of assumptions and employ different definitions of constants characterizing the scattering and absorption properties of the medium. In Kubelka's theory, the constants a, b, and Y are related to the Schuster-Kubelka-Munk (SKM) absorption K and scattering S coefficients as

$$K/S = a-1, \quad SbZ = Y$$

In Chandrasekhar's theory, the true absorption coefficient $\alpha_\nu = \kappa_\nu \rho(1-\bar{\omega}_o)$ and the true scattering coefficient $\sigma_\nu = \kappa_\nu \rho \bar{\omega}_o$. Klier has found the relations between the Chandrasekhar and SKM coefficients

$$\alpha_\nu = \eta K \text{ and } \sigma_\nu = \chi S$$

and tabulated the "correction factors" η and χ as a function of albedo $\bar{\omega}_0$ (10). The ratio (χ/η) is fairly constant and equal, within 1.5%, to (8/3) for values of K/S between 0 and 0.3. The ratio K/S is most easily experimentally determined from the reflectance R_∞ from semi-infinite specimen (Z→∞) as

$$\frac{K}{S} = \frac{(1-R_\infty)^2}{2R_\infty} \tag{5}$$

which is the well known SKM equation.

Klier's contribution consists of demonstrating that the ratio of the true absorption to scattering coefficients (α_ν/σ_ν) can be determined as

$$\frac{\alpha_\nu}{\sigma_\nu} = (\frac{\eta}{\chi}) \frac{(1-R_\infty)^2}{2R_\infty} \tag{6}$$

where η and χ are known numerical factors for each value of R_∞ (or K/S). These corrections become significant for strongly absorbing specimens with (K/S) > 0.3 or R_∞ < 0.5.

It should be emphasized at this point that the expressions (5) and (6) are valid if the medium is a homogeneous isotropic scatterer with no unscattered light propagating in it. There are numerous experimental tests to determine whether these conditions are satisfied (1); in the case they are not satisfied, more complex solutions of the radiative transfer equation have to be employed (9), usually resorting to numerical methods. Experimental input into the theory then involves the phase functions, which have to be determined by a number of tedious measurements of the angular distribution of light intensities. To take advantage of the simplicity of expression (6) for determining (α_ν/σ_ν), an inert white isotropic scatterer of high scattering power may be added to those specimens which themselves do not scatter isotropically. Isotropic scattering is usually produced by randomly oriented irregular particles loosely packed into a powder layer in which the average distances between particles are smaller than the average particle size.

As the scattering coefficients S or σ_ν do not depend on or are only a slight monotonous function of the light frequency ν (1), the variation with ν of the true absorption coefficient K or α_ν determines the spectral structure of (K/S) or (α_ν/σ_ν) and is, except for a multiplicative constant $\frac{1}{S}$ or $\frac{1}{\sigma_\nu}$, a true representation of the energy absorption spectrum of the species contained in the medium. Because catalyst particles of interest usually have a large surface to volume ratio, a substantial part of the observed spectrum may come from surface molecules, complexes, and surface band structure. When the adsorbed species are fairly isolated, the absorption coefficients K or σ_ν are expected to rise

linearly with their concentration, which can be verified by linear
dependence of the experimentally observed

$$F(R_\infty) = (1 - R_\infty)^2 / (2R_\infty),$$

corrected if necessary at high absorption according to equation
(6), against concentration. An example is given in Fig. 1 for
the near-infrared combination band $H_2O(\nu+\delta)$ of water adsorbed on
silica. The integrated intensity ΣI of the absorption band is an
integral of $F(R_\infty)$ over all frequencies ν under the band. Numerous
other examples are given in Kortüm's book ($\underline{1}$) and some additional
examples are shown below for highly absorbing media.

Vibrational Spectra

DRS has been used in both the fundamentals (800–4,000 cm^{-1})
and the overtone-combination (4,000–15,000 cm^{-1}) regions. In the
first region, combination of DRS with rapid scanning Fourier
transform spectrometry with reflected light collection by an
ellipsoidal mirror proved to provide good quality and quantitatively accurate infrared spectra of organic materials mixed with
KBr ($\underline{12}$). The SKM intensities were linear with the concentration of the absorber up to $F(R_\infty)$ = 0.6 and the SKM theory was
considered valid for $F(R_\infty)$ ranging from 0 to 0.6. A direct
observation by DRS of adsorbates and elementary reactions on
catalysts was reported by Kortüm and Delfts ($\underline{13}$), who studied the
interaction of ethylene with the Al_2O_3-SiO_2 cracking catalysts
and concluded that a C_2H_5 surface species is formed from adsorbed
ethylene and hydrogen donated by the surface OH groups. The sensitivity of DRS in infrared fundamentals region was sufficient to
detect the decay of OH intensities upon ethylene adsorption and
to contest earlier transmission IR results ($\underline{14}$) which did not indicate that OH groups were the source of hydrogen for the buildup
of C_2H_5 residues from ethylene. Interesting results were also
obtained by Kortüm and Delfts concerning the adsorbates and products of HCN interaction with various oxide surfaces and with the
cracking catalyst Al_2O_3-SiO_2. The adsorbed monomer, the products
of its decomposition, polymer buildup and dicyan formation were
readily detected by DRS in the IR region. The spectra were not
quantitatively evaluated, however, as much as most IR spectra
obtained by transmission through pellets are not quantitatively
evaluated either. Without doubt the interfacing of both the
reflectance and transmission IR spectrometers with computers will
yield a wealth of quantitative information concerning adsorbates
and their products on catalyst surfaces in the near future.

An example of such a quantitative study is an investigation
of the interactions of surface hydroxyls with adsorbed water on a
variety of silicas ($\underline{15},\underline{16}$). It was established that the surface
hydroxyl groups are hydrogen bond donors, that they form a 1:1
complex with water on non-porous HiSil silicas ($\underline{15}$) and 2:1 com-

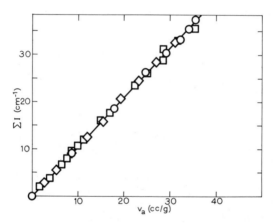

Journal of Physical Chemistry

Figure 1. *The integrated intensities* $\Sigma I = \int_{band} F(R_\infty, v)dv$ *of the* H_2O $(v + \delta)$ *bands*
(○, □) on Na–HiSil(650) and (◇) on HiSil(700) vs. water coverage, v_a. *The tem-*
peratures of heat treatments of these silicas are given in parentheses in degrees Cel-
sius (15).

plex on porous amorphous silicas (16). Since the 1:1 OH...OH$_2$ complex has a binding energy of approximately 6 kcal/mol (17), which is lower than the heat of liquefaction of water, 10.6 kcal/mol, and the 2:1 OH...OH$_2$ complex has a binding energy larger than the heat of liquefaction of water, the non-porous silicas appear hydrophobic whereas the porous silicas are hydrophilic. These properties have important consequences in water clustering and nucleation (18). Surface hydroxyls were also found on aluminum-exchanged fluorine micas (19), the mechanism suggested for their formation being

$$2K^+_{(s)} + Al^{3+}_{(aq)} + H_2O \rightarrow [A\ell(OH)]^{2+}_{(s)} + 2K^+_{(aq)} + H^+_{(aq)}$$

where (s) denotes a cation site in the oxygen six-ring windows of the cleavage plane of mica. These "aluminols" bind strongly and selectively organic phosphates such as nucleotides and nucleic acids by a mechanism depicted in Fig. 2. This property and the fact that the strong Al-O-P bond can be readily cleaved by aqueous fluorides led to the design of efficient separation methods for messenger RNA's (20). It is possible to devise similar methods for anchoring of homogeneous catalysts.

The above results concerning the surface water, OH groups, and their reactivity have been obtained in the overtone and combination bands region, rather than in the fundamentals infrared. The reasons for this are (i) a particularly "clean" distinction between the various vibrational modes of water and of the OH groups (21), and (ii) employment of quartz windows and PbS detectors which permits an easy sample handling and a relatively fast multiscanning over the wavelength range 4,000-15,000 cm^{-1}. The same region proved to be useful for the detection and analysis of organic molecules and functional groups, N$_2$O and CO$_2$ adsorbed on surfaces. A near infrared spectrum of ethylene adsorbed on Mn^{II}A zeolite is shown in Fig. 3. The first two bands at 4450 cm^{-1} and 4660 cm^{-1} are easily identified as the $(\nu_5 + \nu_{12})$ and $(\nu_2 + \nu_9)$ modes by comparison with the earlier observed overtone spectra of ethylene (22), both bands being shifted by -65 cm^{-1} from their gas phase analogues. The second observed set of bands at 5860 cm^{-1} and 6040 cm^{-1} have not been reported earlier but their frequencies are close to double the frequency of the ν_{11}(CH-b$_{3u}$) and ν_9(CH-b$_{2u}$) fundamentals and are therefore assigned the overtone labels $2\nu_{11}$ (5860 cm^{-1}) and $2\nu_9$(6040 cm^{-1}). The structure of the ethylene-Mn^{II} complex is expected to be similar to one determined by Seff for the acetylene-Mn^{II} complex (23). That ethylene is π-bonded to the divalent ions placed in the oxygen six-ring windows of the zeolite has been shown by investigations of the analogous ethylene-Co^{II}A system (24). The binding energy of ethylene was determined to be 17 kcal/mol (24). Ethylene-ion binding in Type A zeolite, and probably in all aluminosilicates in which cations are placed in oxygen six-ring windows, is thus relative-

BINDING OF POLY - A BY THYMIDYLIC ACID ANCHORED ON Al - MICA

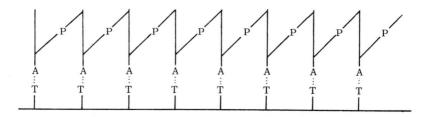

Figure 2.   Thymidylic acid bound through its phosphate group to aluminum-ex-changed mica. The thymidylic acid further adsorbs a poly-A tail of messenger RNA by hydrogen bonding, represented on top. The necessary active center for thymi-dylic acid are surface aluminol groups as discussed in text.

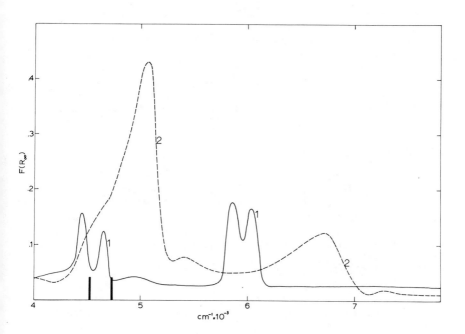

*Figure 3.   Overtone and combination band spectrum of ethylene and water ad-
sorbed on Mn^{II}A zeolite.   The ethylene bands lie close to the gaseous ($v_5 + v_{12}$),
($v_2 + v_9$), $2v_{11}$, and $2v_9$ vibrational modes, indicating that the ethylene molecule has
retained its chemical composition and structural integrity: (——, 1) MnA + ethylene; (– – –, 2) MnA hydrated; (▮) C$_2$H$_4$ (g) bands.*

ly weak but specific and gives rise to well defined structure of
the surface complex.

Because of the small mass of hydrogen, the near-infrared
bands so far discussed were first overtones or simple combination
bands of hydrogen containing species. However, higher overtones
such as the $(\nu_1\nu_2{}^\ell\nu_3)$ = (0 4° 1) band of CO_2 at 4835 cm^{-1} and
more complex combination bands such as the $(\nu_1\nu_2{}^\ell\nu_3)$ = (1 2° 1)
of CO_2 at 4970 cm^{-1} and the $(\nu_1\nu_2{}^\ell\nu_3)$ = (2 0° 1) of CO_2 at 5090
cm^{-1} and of N_2O at 4775 cm^{-1} were observed by DRS in zeolites
where CO_2 and N_2O formed weak ligands to a Cr(II) surface ion (25).

Although DRS has so far yielded smaller volume of data on
vibrational spectra of adsorbates than transmission spectroscopy
of pellets, there are clearly cases where DRS has advantages.
These are summarized as follows: (i) DRS is a good quantitative
technique; (ii) DRS in the near-infrared often allows to resolve
species whose bands overlap in the fundamentals region; (iii) the
sample preparation, handling, and optics enclosing the sample
vessels are often simpler and more expedient than when using
conventional transmission techniques.

Correlation Motion

Given the true vibrational bandshape, the time development of
the molecular reorientation can be determined from Fourier transform of the band intensities onto the time base (26). The ensuing
"correlation function" has the significance of the time development of the transition dipole of the infrared active mode (27).
Where adsorbed molecules are of interest, one can determine
whether they are "fixed", rotate freely, or are hindered in rotation by neighboring ad-molecules or atoms of the surface. A complete temporal description of the molecular motion is in principle
possible by choosing two infrared bands with non-colinear transition dipoles for the determination of the time development of the
two dipole vectors travelling with the molecule. As a matter of
example, the deformation vibration ν_3 of water has a transition
dipole along the molecular axis whereas the antisymmetric stretch
ν_2 perpendicular to it; the time correlation functions of these
two bands contain a complete information about the orientation of
the water molecule at any moment of its tumbling at surfaces, in
gaseous or condensed phases, wherever the vibrational spectrum is
measured. Analysis of correlation motion about one single molecular axis is not as complete but still is of value for determining the rotational mobility of adsorbed molecules. The correlation functions for water in various adsorbates such as on silicas,
micas, and zeolites have been reviewed in ref. 18. The correlation (or reorientational) times can be determined from the correlation functions, or, when the bandshape is known, from the bandwidth. For Gaussian bands, for example, the correlation time is
given as $4\sqrt{\ln 2}/\Delta\omega_{\frac{1}{2}}$, where $\Delta\omega_{\frac{1}{2}}$ is the half-width of the band.
Table I lists some correlation times for water in zeolites, on

silica and mica surfaces, and compares them with those for water
in bulk liquid and in the critical state (28).

TABLE I

Reorientational times τ for water molecules
in various adsorbates and in bulk water.

Sorbent (line)	τ (sec)	Method for determining τ
NaY (5320 cm^{-1})	2.7×10^{-12}	A
NaY (5240 cm^{-1})	1.7×10^{-12}	A
NaY (5120 cm^{-1})	5.3×10^{-13}	A
NaA (5130 cm^{-1})	$(8.2-9.5) \times 10^{-13}$	A
Hi-Sil (ν+δ)	4.0×10^{-14}	B
Fluorine mica (ν+δ)	3.5×10^{-14}	B
Bulk water (ν+δ)	5.0×10^{-14}	B
Critical water (ν+δ)	7.0×10^{-14}	B

A - From half-widths of Gaussian peaks
B - From correlation functions $C(t)$; $C(\tau) = C(0) \cdot e^{-1}$

The relatively long correlation time of the NaY and NaA water
species indicates that intrazeolitic water is rotationally pinned
compared to that on silica and mica surfaces and in bulk water.
On silicas alone, water bound to two surface hydroxyls appears
rotationally less mobile than water bound to a single hydroxyl
(16). On mica surfaces, the reorientation was found fast but
irreversible, indicating angular trapping of this surface species (18).
Many details of fundamental and practical interest can be
obtained from the time correlation functions of adsorbates on
catalysts, similar to the results of studies of adsorbed water
and surface hydroxyls discussed above. To our knowledge, however, no studies of correlation function of adsorbates other than
water have been reported. Perhaps the reason for this is a justified lack of confidence concerning the true handshape as obtained
by traditional methods employing (diffuse) transmission through
pellets. That is not to say that DRS is without problems in the
respect of true bandshapes, involving the possible non-isotropic
scattering where isotropic is assumed, but experimental evidence
is mounting that, with careful application of the radiative
transfer theory, the true lineshapes can indeed be obtained by
spectroscopy of scattering media.

Electronic Spectra of Adsorption Centers and Complexes.

While the purpose of vibrational spectroscopy in mechanistic
studies of adsorption and catalysis is mainly to gain information
about the identity and structure of adsorbed species, *electronic
spectra* are fingerprints of the states of valence electrons in the
catalyst and catalyst-adsorbate complexes. Since most catalysts
are open-shell systems, they have electronic spectra in the UV-
visible-near infrared region; changes of these spectra are readily
observed upon chemisorption, if the catalyst's surface-to-volume
ratio is large. In a number of cases the electronic motion interacts with phonons and vibrations and *vibronic spectra* or vibronic
splitting of electronic bands is observed in the spectral region
between 5,000 cm^{-1} and 40,000 cm^{-1}. Vibronic interactions are of
importance in determining the stability of molecular species and
complexes, and may thus initiate or alter the chemical, including
catalyzed, pathways. It is for this reason that electronic spectra bear perhaps the most direct relation to chemical reactivity
of surfaces. The task to interpret these spectra is formidable,
however, because it involves advanced quantum mechanical analysis
of sizeable molecular-crystal or amorphous cluster systems the
structure of which is often elusive.

A notable exception are chemisorbed complexes in zeolites,
which have been characterized both structurally and spectroscopically, and for which the interpretation of electronic spectra has
met with a considerable success. The reason for the former is the
well-defined, although complex, structure of the zeolite framework
in which the cations are distributed among a few types of available sites; the fortunate circumstance of the latter is that the
interaction between the cations, which act as selective chemisorption centers, and the zeolite framework is primarily only electrostatic. The theory that applies for this case is the ligand field
theory of the ion-molecule complexes usually placed in trigonal
fields of the zeolite cation sites (29). Quantum mechanical exchange interactions with the zeolite framework are justifiably
neglected except for very small effects in resonance energy
transfer (30).

All of the spectral data concerning zeolites have so far been
obtained by DRS. A review of electronic spectroscopy of transition metal ion complexes in Type A zeolites up to 1974 (7) summarizes results concerning the energy levels of the bare sites
occupied by the ions only, and of their 1:1 complexes with water,
olefins, and cyclopropane. Both earlier (8) and more recent (31)
studies also showed that the adsorption process may not stop at
the 1:1 ion-molecule complex but may continue to form intracavi-
tal, multi-liganded complexes, many of which have properties similar to those existing in solutions. The 1:1 complexes have
structures (32) unparallelled in solutions, however, in which the
coordination of the metal ion is low and its energy spectra and
other physical properties such as magnetic moments are pronounced-

ly different from any previously observed in compounds and complexes existing in solutions and crystals.

A particularly detailed description was obtained for complexes of $Co^{II}A$ zeolite with mono-olefins (24). Steric effects due to methyl groups adjacent to the double bond resulted in the ligand strength spectrochemical series

$$\text{ethene} > \text{propene} > \text{cis-butene-2} > \text{trans-butene-2}$$

in which the π-bonded olefins, in the above order, produced decreasing spectral splitting of the 14,000-20,000 cm^{-1} electronic band of the olefin-$Co^{II}A$ complex, as predicted by the ligand field theory for the ^4P state of this low symmetry (C_{2v}) complex (24). The band-splitting in the olefin-$Co^{II}A$ complexes is temperature independent and results from the permanently present low symmetry components of the adsorbed molecules. In related tetrahedral complexes, in which the Co^{II} ions are bound to three skeletal oxygens and one water molecule at the vertices of a tetrahedron, the predicted spectral splitting of the $^4P(^4T_1)$ state gives rise to a symmetric and temperature dependent triplet. Such a behavior is characteristic of the dynamic Jahn-Teller effect within the excited $^4T_1$ state. The transitions and their parameters are shown in Fig. 4. The temperature dependence of the spectral splittings thus allows a distinction to be made between static fields and dynamic distortions and further characterizes the nature of the studied surface complex. A theoretical study of the Jahn-Teller effect in transition metal ion exchanged zeolites showed that off-axial distortions will occur in the bare sites of ions with degenerate grounds states (i.e. Cu^{II}, Cr^{II}, Co^{II}, Ti^{III} of the first transition series) (29). Strong anharmonicities may amplify the effects of vibronic coupling and drive the ions to the proximal framework oxygen, which may result in the destruction of the complex, i.e. in an ion-skeletal chemical reaction. Such an instability-reaction is believed to be the cause of the notorious chemical instability of the $Cu^{II}A$ zeolites.

A chemical reaction can also occur in the adsorbed ligand, and where the original bare ion site has been regenerated by a sequence of reactions, the ion site had performed a function of a catalyst. An example of such a process is the oxidation of CO over the $Cr^{II}A$ zeolite (25). The significance of the $Cr^{II}A$ catalyst rests not in any of its practical or commercial value for CO oxidation but in the fact that every step of oxygen activation for this catalyzed reaction has been characterized spectroscopically, gravimetrically, and by magnetic moment measurements. A mechanism

$$Cr^{II}A + O_{2(g)} \underset{25°C}{\overset{}{\rightleftharpoons}} Cr^{III}A\text{-}O_2^- \xrightarrow{150°C} Cr^{IV}A\text{-}O^{2-} + \tfrac{1}{2}O_{2(g)}$$

$\qquad$ blue $\qquad\qquad\qquad\qquad$ gray $\qquad\qquad\qquad\qquad$ red

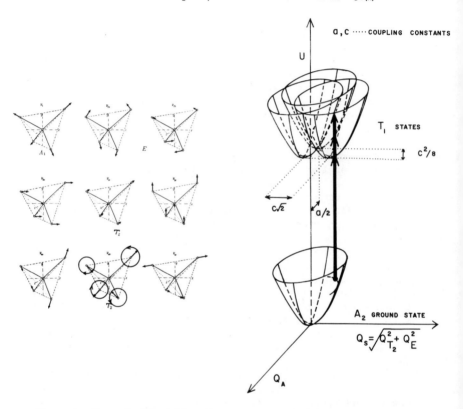

Figure 4. *Dynamic Jahn–Teller effect in the excited states of tetrahedral Co^{II} species. Vibronic interactions of the electronic $^{4}T_{1}$ states with normal modes (left) result in the splitting of the triply degenerate T_{1} state into three intersecting wells (right). Optical transitions are symmetric triplets, the separation of which is temperature dependent: c is the coupling constant for the T_{1} state with the T_{2} normal modes; b the coupling constant of T_{1} with the E normal modes. The resulting "wobbling" motion is indicated in one of the T_{2} normal modes on the left. In zeolites, near-tetrahedral complexes are realized by the four vertices of the tetrahedron being occupied by three skeletal oxygens and a water molecule.*

$$Cr^{IV}_{A-O}{}^{2-} \quad \xrightarrow{\quad CO, 300°C \quad} \quad Cr^{II}A + CO_{2(g)}$$

red   blue

has been well established (25), in conjunction with the earlier
reported reversible oxygen chemisorption on $Cr^{II}A$, (33), depicted
above as the first step. The CO oxidation over $Cr^{II}A$ was also
followed in a continuous flow reactor and reaction observed at
300°C but not at 25°C. It may be concluded that the chemisorbed
oxygen molecular ion O_2^- is not sufficiently activated for oxidiz-
ing CO but the O^{2-} species in the $Cr^{IV}A-O^{2-}$ complex is. The role
of the various oxygen species observed in $Cr^{II}A$ in mild oxidation
reactions has yet to be investigated. There are transition metal
ions on which the O_2^- molecular species is formed but the subsequent O^{2-} is not formed; an example of such a system is the Cu^IY
which is oxidized by molecular oxygen to $Cu^{II}Y-O_2^-$. It is believed that the potential control of catalyzed selective oxida-
tions by the selection of cation and the zeolite matrix has not
been tapped, and although the progress of rational approach to
catalyst selection is slower than that of an empirical approach,
the future advantages of "tailored" over empirical catalysts may
result in a higher selectivity of the former. For catalyst
tailoring a detailed spectral description such as one in the few
examples given above is of substantial benefit because each piece
of information on the electronic structure of adsorbed complexes
generates new ideas for preparations of new effective catalysts
guiding a given reaction to desired products.

Photoluminescence

Light energy absorbed in catalyst particles and surfaces may
be re-emitted, in part, at a different wavelength (or different
wavelengths). When this occurs in scattering media, the emission
coefficient j_ν of equation (2) will have two contributions:
light $j_\nu^{(s)}$ scattered into the direction μ from all other directions without a change of frequency ν, and light $j_\nu^{(e)}$ emitted into the direction μ by photoluminescence. In previous paragraphs
were discussed cases in which $j_\nu^{(e)}$ is equal to zero. The complete radiative transfer equation that includes photoluminescence
reads as follows:

$$- \frac{dI_\nu}{\kappa_\nu \rho ds} = I_\nu - \frac{1}{\kappa_\nu} (j_\nu^{(s)} + j_\nu^{(e)}) \tag{7}$$

Here $j_\nu^{(s)}$ takes any of the forms discussed earlier and $j_\nu^{(e)}$, in
the case that emission is isotropic and proportional to excitation light intensity at some other frequency ν', becomes

$$j_\nu^{(e)} = c_{\nu\nu'} \kappa_{\nu'} \int p_{\nu'}(\mu') I_{\nu'}(\mu', \tau) \, d\mu' \tag{8}$$

An additional radiative transfer equation

$$- \frac{dI_{\nu'}}{\kappa_{\nu'} \rho ds} = I_{\nu'} - \frac{1}{\kappa_{\nu'}} j_{\nu'}^{(s)} \tag{9}$$

holds for the distribution of the excitation intensity $I_{\nu'}$.
Equations (7) - (9) specify the radiative field in the presence of
photoluminescence, providing that the phase function $p_{\nu'}$ and the
boundary conditions are known. In the case of isotropic scatter-
ing in plane parallel media, $p_{\nu'} = \bar{\omega}_o(\nu')$, equation (9) has the
hyperbolic solutions outlined in ref. 10 and the distribution of
the "emitted" intensity $I_{\nu}(\mu)$ becomes known by solving equations
(7) and (8). An undetermined constant is the quantum efficiency
$c_{\nu\nu'}$ of energy transfer from the excitation frequency ν' to the
emission frequency ν. A straightforward method for determining
$c_{\nu\nu'}$ is the use of internal luminescent standard in the scattering medium under investigation. Thus the photoluminescence quantum efficiencies can be determined in turbid media of known scattering properties as accurately as in optically homogeneous media
such as solutions and crystals. The frequency (ν) dependence of
κ_{ν} is known as *absorption spectrum*, of $j_{\nu}^{(e)}$ as *emission spectrum*,
and of $c_{\nu'\nu}$ as *excitation spectrum* of emission at frequency ν'.
In addition to these three kinds of spectra, the photoluminescence
decay time can be measured after pulsed or step illumination.
     Measurements of absorption, excitation, and emission spectra
as well as of decay times have been made for zeolite specimens
containing univalent copper ions. After initial reports of Cu^I
photoluminescence in Type Y zeolite by Maxwell and Drent (34),
Texter et al. (35) offered a first interpretation of the transi-
tion involved in the emission as a $^3E^-[3d^9\ 4(sp)] \rightarrow {}^1A_1[3d^{10}]$,
where the $^3E^-$ state is the lowest lying, Jahn-Teller split triplet
state of the excited configuration 3d 4(sp) of the Cu^I ion in the
trigonal (C_{3v}) zeolite field. Strome and Klier (36) demonstrated
that new emission lines were produced by CO adsorption-desorption
and interpreted this result as a migration of the Cu^I ions from
the original SI' positions to the surface SII and SII' positions.
Strome (30) carried out a detailed study of luminescence life-
times and of energy transfer in Cu^IY zeolites co-exchanged with
Ni^{II}, Co^{II}, and Mn^{II} ions and determined the quantum efficiencies
of the green Cu^IY emission in the presence of the other colored
ions. The experimental quantum efficiencies are compared in Fig.
5 with those derived from a model using statistical ion distribution and the Foerster-Dexter theory of resonance transfer. The
agreement between the experimental and theoretical values is
excellent and shows that photoemission data obtained by DRS are
amenable to valid and rigorous theoretical interpretation which
can be used for the determination of electronic interactions between the emitters and acceptors across relatively large distances. The basic component of the transition probability for exci-

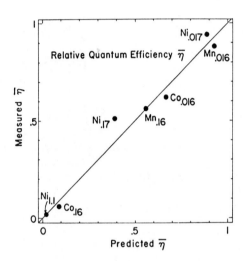

Figure 5. Relative quantum efficiencies of photoluminescence in Cu^{I}Y zeolites containing co-exchanged ions. The predicted quantum efficiencies were calculated from overlaps of the emission band of Cu^I with the absorption band of the co-exchanged ion using the Foerster–Dexter resonance transfer theory.

tation energy transfer between the emitter (Cu^I) and the acceptor
center is the overlap of the emission band with the absorption
spectrum of the acceptor (Ni^{II}, Co^{II}, Mn^{II} in this case) (37).
Similar energy transfer was observed between the luminescing Cu^I
ions and non-luminescing Cu^{II} ions that were produced in the Cu^IY
zeolite by partial oxygen chemisorption $Cu^IY + O_{2(g)} \rightarrow Cu^{II}Y\text{-}O_2^-$.
The Cu^{II} ions have a weak absorption spectrum that partially over-
laps with the emission band of Cu^I, resulting in resonant energy
transfer.   In fact the time course of oxygen chemisorption could
be followed by monitoring the Cu^I photoluminescence quantum efficiency with the time of exposure of Cu^IY to oxygen.

     Although applications of photoemission techniques in surface
chemistry and catalysis are but a few, their sensitivity, which
is orders of magnitude higher than that of adsorption measurements, may lead to future investigations of very small surfaces or
adsorbed complexes in very small concentrations.   The technique
is not universal, however, because relatively few surface species
will display photoluminescence.

Electronic Spectra of Bulk Catalyst Particles

     In two instances are the electronic spectra of the bulk of
the catalyst particles of interest in catalysis research: first,
when chemisorption gives rise to electron exchange that extends
to large distances into the solid and second, when various components of a multiphase catalyst interact so as to dope one phase
with the chemical elements of another, resulting in new, enhanced,
or reduced activity of the catalyst.

     An example of the former effect is the well-known "blacken-
ing" of nickel oxide upon oxygen chemisorption: the pure green
stoichiometric NiO has a DRS spectrum characteristic of Ni^{II} ions
in the octahedral environment of the nearest neighbor oxygen
anions of the rock-salt lattice of NiO while the oxygen-covered
NiO displays an additional intense charge transfer band at 550 nm
which eventually overlaps the whole visible spectrum (38).   This
spectral change results from a transfer of electrons from the
surface nickel ions to chemisorbed oxygen, $\frac{1}{2} O_{2(g)} + Ni^{II} \rightarrow O^-_{(ads)}$
$-Ni^{III}$,   and a subsequent exchange of the excess positive charge
of Ni^{III} with Ni^{II} of the bulk.   The latter process can be
observed as an increase of the p-conductivity of the NiO upon
oxygen exposure (39) and it is the $Ni^{III}_{(1)} + Ni^{II}_{(2)} \rightleftharpoons Ni^{II}_{(1)} + Ni^{III}_{(2)}$
transition that gives rise to the optical charge transfer band at
550 nm.   The process can be reversed by a reaction of the chemisorbed oxygen with oxidizeable gases.   For example, CO will
react at room temperature as $CO_{(g)} + O^-\text{-}Ni^{III} \rightarrow CO_{2(ads)} + Ni^{II}$
(40), restoring the spectrum and green color of the stoichiomet-
ric nickel oxide and lowering the p-conductivity to the original
level.   The ease with which nickel oxide chemisorbs oxygen and
$O^-\text{-}Ni^{III}$ reacts with CO would make nickel oxide an excellent
low temperature catalyst for CO oxidation, were the whole process

not limited by the desorption of CO_2. The merit of spectroscopic
techniques in this case is to resolve the reaction mechanism
encompassing the electron exchange with the catalyst and to find
the rate determining step of the reaction.

A good example of an application of DRS to the determination
of the active component in a multiphase catalyst is the recent
study of the $Cu/ZnO/Al_2O_3$ and $Cu/ZnO/Cr_2O_3$ methanol synthesis
catalysts (41-43). While none of the components alone is a catalyst for the synthesis at low (less than 100 atm) pressures, the
ternary systems are highly active and selective catalysts at
those pressures. The activity resides in the bi-phase Cu/ZnO
system (41) which was studied by DRS in some detail (43). It was
established that, although zinc oxide was present in its ordinary
wurtzite crystal form, its characteristic optical absorption edge
at 25,800 cm^{-1} was completely missing in the most active catalysts, and a new band appeared in the visible at 17,000 cm^{-1} due
to copper dissolved in the zinc oxide lattice. These findings
were confirmed by analytical electron microscopy which determined
the amount of dissolved copper to be up to 16% in the zinc oxide
phase (42). The genesis of the active methanol catalyst was
followed from precipitate precursors through the calcination and
reduction stages, with subsequent analysis by various methods
including DRS, Auger/XPS (X-ray photoelectron spectroscopy), STEM
(scanning transmission electron microscopy), X-ray diffraction,
TEM (transmission electron microscopy), surface area and chemisorption methods, and pore distribution determinations. A
description of the catalyst as complete as obtained in these
studies would not have been possible without employing a great
many techniques of catalyst characterization. Among these, however, DRS stands out as little time consuming, inexpensive, sensitive to electronic interactions among the catalyst components,
and having high spectral resolution. For these reasons, DRS is
likely to remain an established technique for characterization of
dispersed catalysts and, as was shown in earlier discussion, also
their surfaces. The present limitation of the latter stems from
the limitations of the theory of surfaces. It is hoped, however,
that with the advances in structural information on a molecular
scale, such as from high resolution STEM, the theory will anchor
itself on reliable and realistic models and provide background for interpretation of DRS in highly interacting systems.

Abstract

Diffuse Reflectance Spectroscopy (DRS) is suited for the
study of real catalysts as it measures and interprets light
intensity loss spectra in absorbing and scattering specimens.
DRS has been applied both to the analysis of vibrational spectra of
surface species in the fundamental, overtone, and combination
band regions, and to the determination of time correlation motion
of adsorbed molecules by Fourier inversion of the spectra onto

the time base. In addition, electronic spectra of the chemisorption centers have been studied prior and during chemisorption which, in combination with the vibrational spectra, resulted in a nearly complete determination of the structure of adsorbed complexes between hydrocarbons, oxygen, carbon monoxide, water, metal ions, and aluminosilicate surfaces. As the recent research results indicated the importance of interactions among the catalyst components in determining its selectivity and activity, DRS has proved a powerful tool assisting the analysis of chemical and physical interactions in multicomponent catalysts by other techniques such as scanning transmission electron microscopy. Examples are given from the analysis of the copper-zinc oxide based methanol and water gas shift catalysts.

Combined with photoemission, DRS provides quantitative data on excitation-luminescence behavior of powdered specimens which can be used to determine photoluminescence quantum efficiencies and the extent of resonant energy transfer among the bulk and surface activators and sensitizers.

Literature Cited

1. Kortüm, G., "Reflectance Spectroscopy. Principles, Methods, and Applications". Springer-Verlag, Berlin and New York, 1969.
2. Wendlandt, W.W., and Hecht, H.G., "Reflectance Spectroscopy". Wiley (Interscience), New York, 1966.
3. Wendlandt, W.W., ed., "Modern Aspects of Reflectance Spectroscopy", Plenum, New York, 1968.
4. Frei, R.W., and MacNeil, J.D., "Diffuse Reflectance Spectroscopy in Environmental Problem Solving". CRC Press, Cleveland, Ohio, 1973.
5. Delgass, W.N., Haller, G.L., Kellerman, R., and Lunsford, J.H., "Spectroscopy in Heterogeneous Catalysis". Academic Press, New York, 1979.
6. Jones, C.E., and Klier, K., Annual Revs. Mater. Sci. 2, 1 (1972).
7. Kellerman, R., and Klier, K., Surface and Defect Properties of Solids (Chem. Soc. London) 4, 1 (1975).
8. Klier, K., Catalysis Revs. 1, 207 (1967).
9. Chandrasekhar, S., "Radiative Transfer". Dover, New York, 1960.
10. Klier, K., J. Opt. Soc. Am. 62, 882 (1972).
11. Kubelka, P., J. Opt. Soc. Am. 38, 448 (1948).
12. Fuller, M.P., and Griffiths, P.R., Anal. Chem. 50, 1906 (1978).
13. Kortüm, G., and Delfts, H., Spectrochimica Acta 20, 405 (1964).
14. Lucchesi, P.J., Charter, J.L., and Yates, D.J.C., J. Phys. Chem. 77, 1457 (1962).

15. Klier, K., Shen, J.H., and Zettlemoyer, A.C., <u>J. Phys. Chem.</u> <u>77</u>, 1458 (1973).
16. Shen, J.H., and Klier, K., <u>J. Colloid Interface Sci.</u>, in press.
17. Bassett, D.R., Boucher, E.A., and Zettlemoyer, A.C., <u>J. Colloid Interface Sci.</u> <u>34</u>, 3 (1970).
18. Klier, K., and Zettlemoyer, A.C., <u>J. Colloid Interface Sci.</u> <u>58</u>, 216 (1977).
19. Huang, S.-D., Pulkrabek, P., and Klier, K., <u>J. Colloid Interface Sci.</u> <u>65</u>, 583 (1978).
20. Pulkrabek, P., Klier, K., and Grunberger, D., <u>Anal. Biochem.</u> <u>68</u>, 26 (1975).
21. The following IR species of water and surface hydroxyls have been observed:

$H_2O(\nu_2 + \nu_3)$ around 5300 cm^{-1}, $H_2O(\nu_1 + \nu_2)$ at 7150 cm^{-1}, SiOH(2ν) at 7300 cm^{-1}, and SiOH(2$\nu + \delta$) at 8100 cm^{-1}.

22. Herzberg, G., "Molecular Spectra and Molecular Structure II. Infrared and Raman Spectra of Polyatomic Molecules". Van Nostrand Reinhold Co., New York 1945, p. 326.
23. Riley, P.E., and Seff, K., <u>J. Amer. Chem. Soc.</u> <u>95</u>, 8180 (1973).
24. Klier, K., Kellerman, R., and Hutta, P.J., <u>J. Chem. Phys.</u> <u>61</u>, 4224 (1974).
25. Kellerman, R., and Klier, K., Molecular Sieves-II, ACS Symposium Series <u>40</u>, 120 (1977).
26. Gordon, R.G., <u>J. Chem. Phys.</u> <u>43</u>, 1307 (1965).
27. Klier, K., <u>J. Chem. Phys.</u> <u>58</u>, 737 (1973).
28. Shen, J.H., Zettlemoyer, A.C., and Klier, K., <u>J. Phys. Chem.</u>, in press.
29. Klier, K., Hutta, P.H., and Kellerman, R., Molecular Sieves-II, ACS Symposium Series <u>40</u>, 108 (1977).
30. Strome, D.H., Dissertation, Lehigh University, 1977.
31. Lunsford, J.H., Molecular Sieves-II, ACS Symposium Series <u>40</u>, 473 (1977).
32. Seff, K., <u>Accounts of Chemical Research</u> <u>9</u>, 121 (1976).
33. Kellerman, R., Hutta, P.J., and Klier, K., <u>J. Am. Chem. Soc.</u> <u>96</u>, 5946 (1974).
34. Maxwell, I.E., and Drent, E., <u>J. Catal.</u> <u>41</u>, 412 (1976).
35. Texter, J., Strome, D.H., Herman, R.G., and Klier, K., <u>J. Phys. Chem.</u> <u>81</u>, 333 (1977).
36. Strome, D.H., and Klier, K., <u>J. Phys. Chem.</u>, in press.
37. Dexter, D.L., <u>J. Chem. Phys.</u> <u>21</u>, 836 (1953).
    Förster, Th., <u>Ann. Phys.</u> <u>2</u>, 55 (1948).
    Dexter, D.L., Förster, Th., and Knox, R.S., <u>Phys. Stat. Sol.</u> <u>34</u>, K159 (1969).
38. Klier, K., <u>Kinetics and Catalysis</u>, <u>3</u>, 65 (1962).
39. Kuchynka, K., and Klier, K., <u>Coll. Czech. Chem. Communications</u> <u>28</u>, 148 (1963).

40. Klier, K., and Jiratova, M., Proceedings of the Third Inter-
     national Congress on Catalysis, North-Holland Pub. Co.,
     Amsterdam 1965, p. 763.
41. Herman, R.G., Klier, K., Simmons, G.W., Finn, B.P., Bulko,
     J.B., and Kobylinski, T.P., J. Catal. 56, 407 (1979).
42. Mehta, S., Simmons, G.W., Klier, K., and Herman, R.G., J.
     Catal. 57, 339 (1979).
43. Bulko, J.B., Herman, R.G., Klier, K., and Simmons, G.W.,
     J. Phys. Chem., 83, 3118 (1979).

RECEIVED June 3, 1980.

The Application of High Resolution Electron Energy Loss Spectroscopy to the Characterization of Adsorbed Molecules on Rhodium Single Crystal Surfaces

L. H. DUBOIS and G. A. SOMORJAI

Materials and Molecular Research Division, Lawrence Berkeley Laboratory and Department of Chemistry, University of California, Berkeley, CA 94720

The scattering of low energy electrons by metal surfaces has been studied for many years now (1, 2). The electron's ease of generation and detection and high surface sensitivity (low penetration depth) make it an ideal probe for surface scientists (1, 2). The impinging electron can interact with the surface in basically two ways: it can either elastically reflect (or diffract) from the surface without losing energy or lose a portion of it's incident energy and inelastically scatter. In this paper we will be concerned with only one of many possible inelastic scattering processes: the loss of the electron's energy to the vibrational modes of atoms and molecules chemisorbed on the surface. This technique is known as high resolution electron energy loss spectroscopy (or ELS, EELS, HRELS, HREELS, etc.)

Overview

The Scattering Process. Evans and Mills (3) and Ibach (4) showed that the photon-surface oscillating dipole and the long range electron-surface oscillating dipole interactions are similar and therefore, to first order, infrared spectroscopy and high resolution ELS provide the same information. Furthermore, due both to the long range nature of this interaction (3, 4) and to the large dielectric constant of the substrate, the same selection rules should apply. Specifically, only vibrational modes which contain an oscillating dipole moment with a component perpendicular to the surface can be excited (3, 4). Sokcevic et al. (5, 6, 7) showed that for ELS this normal dipole selection rule can break down under certain conditions, however. Furthermore, for the case of short range "impact" scattering, both theory and experiment indicate that all possible normal mode vibrations are observed (8, 9, 10). A more detailed discussion of the theory of the inelastic scattering of low energy electrons by surface vibrational modes is presented elsewhere in this volume (11).

0-8412-0585-X/80/47-137-163$07.00/0
© 1980 American Chemical Society

The beginning of high resolution electron energy loss spectroscopy can be traced to the pioneering work of F. M. Propst and T. C. Piper in 1967 (12). Credit for the expansion of this technique, however, must be given to Harold Ibach and his graduate students (2, 13) who played a key role in both the advancement of spectrometer design and in the performance of outstanding experiments. At present there are almost a dozen working systems throughout the world and more are being built every month.

To study this very low energy inelastic scattering process in the limit of a long range electron-dipole interaction, a collimated beam of monochromatic electrons is incident on a surface and the energy distribution of the specularly reflected beam is recorded. A typical spectrometer for this type of experiment is shown in Figure 1. Electrons are emitted by a hot tungsten filament and are focused onto the slit of a monochromator by an electrostatic lens system. In our spectrometer the monochromator consists of a 127° cylindrical sector although 180° hemispheres (14, 15, 16) and cylindrical mirror analyzers (17, 18) have also been successfully used as energy dispersing elements. For an excellent review of the various types of electrostatic energy analyzers and their advantages and disadvantages, see reference 19. A monoenergetic beam of electrons is then focussed onto the sample by another series of lenses. A 0.1 to 1 nanoamp beam of electrons reaches the crystal with an incident energy between 2 and 10 eV. The specularly reflected electrons are collected by a pair of lenses and focussed onto the slit of the analyzer. The energy analysis of these inelastically scattered electrons is carried out by a cylindrical sector identical to the monochromator. The electrons are finally detected by a channeltron electron multiplier and the signal is amplified, counted and recorded outside of the vacuum chamber. A typical specularly reflected beam has an intensity of 10^5 to 10^6 electrons per second in the elastic channel and a full width at half maximum between 7 and 10 meV ($60-80$ cm^{-1}; 1 meV = 8.065 cm^{-1}). Scattering into inelastic channels is between 10 and 1000 electrons per second. In our case the spectrometer is rotatable so that possible angular effects can also be studied. This becomes important for the study of vibrational excitation by short range "impact" scattering (8, 9, 10).

The Advantages of High Resolution ELS. The advantages of this technique for studying the vibrational spectra of chemisorbed molecules are numerous: a) The entire infrared region of the spectrum (from 300 to 4000 cm^{-1}) can be scanned in less than twenty minutes without changing windows or prisms. b) These studies can easily be extended to the visible portion of the electromagnetic spectrum (20). c) For strong scatterers, high resolution ELS is sensitive to less than 0.1% of a monolayer (13), far more sensitive than most other techniques. d) Because of both this surface sensitivity and the nature of the inelastic

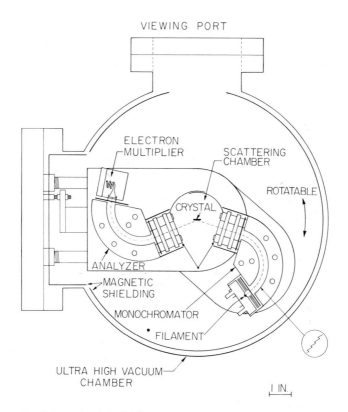

Figure 1.   Schematic of the high resolution electron energy loss spectrometer used in these studies. The dispersive elements are 127° cylindrical sectors.

scattering process ($\underline{3}$, $\underline{4}$, $\underline{5}$, $\underline{6}$, $\underline{7}$), single crystal metal surfaces make ideal targets.   Thus, studies can be carried out on clean, well characterized substrates. Furthermore, e) a variety of complementary surface sensitive probes (such as low energy electron diffraction (LEED), Auger electron spectroscopy (AES), thermal desorption mass spectrometry (TDS) and X-ray and ultra-violet photoelectron spectroscopy (XPS and UPS)) can now be applied to the study of a given chemisorption system. Through this combination of techniques located in a single ultra-high vacuum chamber, a complete picture of the structure, bonding and reactivity of adsorbed monolayers can be obtained. f) Unlike some other techniques, both disordered ($\underline{21}$) and optically rough ($\underline{22}$) surfaces can readily be studied. g) Also, because of this surface sensitivity ELS can be used to detect hydrogen both through its vibration against the substrate and against other adsorbed atoms. Finally, h) due to the low incident beam energies and beam currents, high resolution electron energy loss spectroscopy is a non-destructive technique which can be used to probe the structure of weakly absorbed molecules.

While discussing the advantages of high resolution ELS, mention should be made of two of it's major disadvantages. First, by optical standards the terms "high resolution" are a misnomer since the resolution is limited, at present, to about 60 cm^{-1} (the full width at half maximum of the elastic scattering peak). Peak assignments can be made much more accurately (± 5 cm^{-1}), however. This limits the use of isotopic substitution and the analysis of closely spaced vibrational modes. The second major drawback is that the maximum pressure under which experiments can be carried out is approximately 5×10^{-5} torr due to electron-gas collisions inside the spectrometer. High pressure catalytic reactions and chemisorption at the solid-liquid interface cannot be readily studied. Nevertheless, a tremendous number of studies on the adsorption of atoms, diatomic molecules and large hydro-carbons on transition metal surfaces have been performed.

The Application of High Resolution ELS. In the remainder of this paper we wish to show the power of high resolution ELS by discussing in detail its application to two chemisorption systems. Specifically, we will deal with the adsorption of carbon monoxide on the Rh(111) single crystal surface ($\underline{23}$) and the adsorption and subsequent reactions of acetylene and ethylene on Rh(111) ($\underline{24}$). The vibrational spectra of chemisorbed CO indicate two distinct binding sites (most likely atop and bridged) whose relative populations and vibrational frequencies are determined as a function of both the substrate temperature and background pressure (coverage). Thermal desorption mass spectroscopic measurements show the bridge bonded CO to have an approximately 4 kcal/mole lower binding energy to the surface than the species located in the atop site and therefore this species can be selectively removed from the rhodium crystal.

Surface pretreatment also has a marked effect on CO adsorption: oxygen and carbon both inhibit carbon monoxide chemisorption and weaken the metal-adsorbate bond strength. Co-adsorbed hydrogen has no observable effects.

Below 270 K vibrational spectroscopic measurements indicate that acetylene chemisorbs on Rh(111) with its C-C bond oriented parallel to the surface forming an approximately sp^2 hybridized species. LEED investigations show that both C_2H_2 and C_2H_4 form metastable (2x2) structures on this surface below 270 K. An irreversible order-order transformation occurs between 270 and 300 K to a stable c(4x2) hydrocarbon overlayer in both cases. High resolution ELS and TDS studies show that the stable species formed from both molecules are identical. Hydrogen addition to chemisorbed acetylene is necessary to complete this conversion, however. The structure of the adsorbed ethylene species does not change during this transformation. This stable hydrocarbon species is identical to the hydrocarbon species formed from the chemisorption of either C_2H_4 or C_2H_2 and hydrogen on Pt(111) above 300 K. Decomposition of these molecules to surface CH (CD) species occurs on Rh(111) above ~420 K.

Experimental

Experiments are carried out in an all stainless steel ultrahigh vacuum (UHV) chamber built in two levels. The upper portion contains the standard single crystal surface analysis equipment (four grid LEED/Auger optics, glancing incidence electron gun and quadrapole mass spectrometer). After dosing, the samples are lowered into the high resolution electron energy loss spectrometer by an extended travel precision manipulator. The spectrometer is described in detail above (see Figure 1). The design is similar to that of Froitzheim et al. (25). In the present series of experiments the angle of incidence is fixed at 70° to the surface normal and electrons are collected in the specular direction. The elastic scattering peak has a full width at half maximum between 60 and 100 cm^{-1} and a maximum intensity of 1 x 10^5 counts per second. The vacuum chamber is lined with layers of μ metal and silicon-iron shielding to reduce stray magnetic fields. The upper and lower levels are also separated by μ metal. The base pressure in the system is maintained at 1 x 10^{-10} torr with two sputter ion pumps and a titanium sublimation pump.

This combination of techniques allows us to determine the structure of the adsorbed species while on the metal surface and after desorption into the gas phase. Furthermore, molecular rearrangements in the adsorbed overlayer as a function of both the substrate temperature and background pressure can be studied.

The procedures for sample preparation, mounting and cleaning have been described previously (26). Briefly, the rhodium single crystal rod was oriented to ±1/2° using X-ray back reflection and a 1 mm. thick disc was cut by spark erosion. After mechanical

polishing the sample was spot welded to etched tantalum foil and
mounted inside the UHV chamber. The Rh(111) crystal was cleaned
by a combination of argon ion bombardment (1000-2000 eV) followed
by annealing in vacuum (~900 K) and O_2/H_2 cycles to remove carbon,
sulfur and boron.

Gas adsorption is studied at pressures between 5 x 10^{-9}
and 1 x 10^{-5} torr and at temperatures between 210 and 530 K.
Surface structures are observed both with increasing exposure
and after the gas is pumped away. Neither gas exposures nor
background pressures are corrected for ion gauge sensitivity.

The Chemisorption of CO on Rhodium

Background. The bonding of carbon monoxide to transition metals is one of the most extensively studied chemisorption
systems in surface science (27, 28). The interaction of CO with
metal atoms or with clusters of metal atoms is also well studied
by inorganic chemists (29, 30, 31). As a result of these detailed
investigations of CO, comparisons between surfaces and metal
cluster carbonyls can now be made (32, 33, 34). The bonding of
carbon monoxide to rhodium is of special interest to us since
this metal catalyzes the hydrogenation of CO in both heterogeneous
(35, 36) and homogeneous media (37). Because of this importance
we chose to explore the high resolution electron energy loss
spectrum of carbon monoxide chemisorbed on a Rh(111) single
crystal surface (23).

Much effort has gone into understanding the vibrational
spectrum of CO chemisorbed on rhodium (38). More than 20 years
ago Yang and Garland performed the first infrared studies using
highly dispersed rhodium particles supported on an inert alumina
substrate (39). They provided convincing evidence for a species
of the form $\overline{Rh}(CO)_2$, a gem dicarbonyl, whose IR spectrum showed
a doublet at 2095 and 2027 cm^{-1}. The presence of linear (~2060
cm^{-1}) and bridge bonded (1925 cm^{-1}) forms were also demonstrated.
More recent investigations (38, 40, 41, 42, 43, 44) agree
extremely well with these early experiments except that the C = 0
vibration of the multiply coordinated species was found to lie
near 1860 cm^{-1} in most cases. Supported rhodium cluster carbonyls
of known molecular structure have also been studied and analogous
stretching frequencies in the 1800 to 2100 cm^{-1} region were
reported (45,46). Substrate absorption below 1000 cm^{-1} masked all
Rh-CO vibrations. Both transmission and reflection IR studies
employing evaporated rhodium films yield similar results (38).
Due to the high density of Rh atoms, no species of the form
$Rh(CO)_2$ were formed, however. This is expected to be the case
on the (111) surface as well. Weak absorptions between 400 and
575 cm^{-1} were seen and are indicative of metal-absorbate stretching and bending vibrations. Inelastic electron tunneling
spectroscopic (IETS) measurements on alumina supported rhodium
particles (47, 48, 49) add little new structural information

except that peaks in the 400-600 cm^{-1} region have been definitely assigned to Rh-CO bending and stretching modes.

CO Chemisorption on Clean Rh(111) (23). The vibrational spectra of carbon monoxide chemisorbed on Rh(111) at 300 K as a function of exposure are shown in Figure 2. At very low exposures (less than 0.1 L; 1 L = 1 Langmuir = 10^{-6} torr · sec) only one peak at 1990 cm^{-1} is observed in the C ≡ 0 stretching region and no ordered LEED pattern is found. By comparison with the infrared spectra of relevant organorhodium compounds (50, 51) and with matrix isolated metal carbonyls (52), one can assign this loss to the carbon-oxygen stretching vibration of a linearly bonded species. This peak shifts to higher frequency as the coverage is increased. Possible causes for this include local field effects (53, 54), vibrational coupling (54), dipole-dipole interactions (55) or simply a decrease in the total metal-carbon backbonding due to the increased number of adsorbate molecules (56). Currently one lacks a complete understanding of the changes of the forces involved and the individual effects of each perturbation cannot be completely decoupled. Recent infrared studies on single crystal platinum (57) and copper (58) surfaces have attempted to sort out these effects, but due to the extent of the shift here, a combination of all of these mechanisms is most likely present. A result of these effects is a decrease in the rhodium-carbon stretching frequency and possibly a weakening of the metal adsorbate bond. Figure 2 clearly shows a shift in the rhodium-carbon stretching vibration for this linearly bonded species from 480 cm^{-1} to lower frequency with increasing CO exposure. This is consistent with the weakening of the metal-adsorbate bond observed with increasing gas dosage in CO thermal desorption spectra from the Rh(111) surface (23, 26, 59). No other vibrations corresponding to Rh-C ≡ 0 bending modes were observed in the specular direction and by invoking the normal dipole selection rule (3, 4) we conclude that the carbon-oxygen bond is oriented perpendicular to the surface.

At larger than 0.4 to 0.5 L CO exposures a small shoulder near 1870 cm^{-1} appears. Again by comparison with relevant model compounds (50, 51, 52), one can assign this peak to the carbon-oxygen stretch of a bridge bonded species. Unlike the loss near 2000 cm^{-1}, this peak grows at essentially constant frequency, never varying more than ±5 cm^{-1}. By a CO exposure of 1.0 L the rhodium-carbon stretch has significantly broadened. The new low frequency shoulder appearing slightly above 400 cm^{-1} corresponds to the metal-carbon stretch of the bridge bonded species. This weaker bond to the substrate for the new species can be correlated with a low temperature desorption peak appearing at high CO exposures in the TDS spectra of reference 23. These measurements show the bridge bonded CO to have an approximately 4 kcal/mole lower binding energy to the surface than the

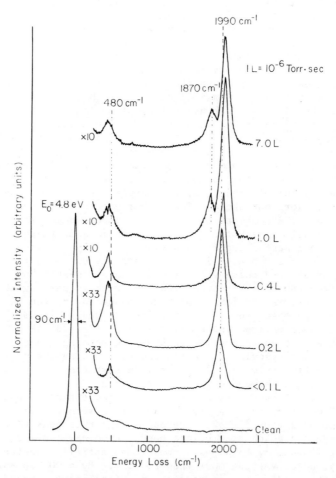

Figure 2.   Vibrational spectra of CO chemisorbed on an initially clean Rh(III) single crystal surface at 300 K as a function of gas exposure.   Note the shift in both the 480 and 1990 cm⁻¹ losses with increasing surface coverage.

species located in the atop site. Again the bridge bonded species is oriented perpendicular to the surface since no bending or asymmetic stretching modes are observed in the specular direction.

The vibrational spectra of CO chemisorbed on Rh(111) as a function of background pressure at 300 K are shown in Figure 3. Once again the carbon-oxygen stretch for the atop site continues to shift to higher frequency as a function of coverage and reaches a limiting value of 2060 to 2070 cm^{-1}. The rhodium-carbon stretch of the linear species simultaneously decreases to 420 cm^{-1}. The 1870 cm^{-1} loss due to the bridge bonded species remains at a constant frequency with increasing coverage, however. The presence of gem dicarbonyl species cannot be ruled out here due to the limited resolution of ELS, but seem unlikely because of the high density of metal atoms on the (111) surface (38) that would lead to extreme crowding of CO molecules in the dicarbonyl configuration.

The chemisorption of carbon monoxide on Rh(111) is completely reversible. As the background CO in Figure 3 is pumped away the carbon-oxygen stretching vibration for the more weakly bonded bridged species decreases in intensity and the metal-carbon and carbon-oxygen stretching vibrations for the atop site shift back into their original positions. The bridge bonded species can be selectively removed from the substrate by slowly heating the crystal to approximately 360 K in vacuum (23). Finally, no CO decomposition is detected under any of the conditions employed in our experiments ($p \leqslant 1 \times 10^{-5}$ torr CO, $T \leqslant 600$ K).

By combining the present ELS studies with previous TDS (23, 26, 59) and LEED (23, 59) experiments we can now present a fairly complete picture of CO chemisorption on Rh(111). At very low exposures a single species is present on the surface located in an atop site (ν_{Rh-CO} = 480 cm^{-1}, $\nu_{C\equiv O}$ = 1990 cm^{-1}). As the coverage increases, the bonding to the surface becomes weaker (ν_{Rh-C} decreases, $\nu_{C\equiv O}$ increases, TDS peak maximum shifts to lower temperatures) (23, 26, 59). This process continues until an approximately 0.5 L exposure where a ($\sqrt{3} \times \sqrt{3}$)R30° LEED pattern is seen and all of the adsorbed CO molecules are linearly bounded to individual rhodium atoms (Figure 4a). This has now been confirmed by a low energy electron diffraction structure analysis (60). Above this coverage a second C = O stretching vibration corresponding to a bridge bonded species is observed ($\nu_{Rh-CO} \approx$ 400 cm^{-1}, $\nu_{C=O}$ = 1870 cm^{-1}). A "split" (2x2) LEED pattern is seen indicating a loosely packed hexagonal overlayer of adsorbate molecules. This overlayer structure compresses upon further CO exposure. Throughout this intermediate coverage regime there is a mixed layer of atop and bridge bonded CO species and we see a continuous growth of all ELS peaks and a shift in the loss above 2000 cm^{-1}. Two peaks are also visible in the TDS spectra with the bridge bonded carbon monoxide having an approximately 4 kcal/mole lower binding energy to the surface than the species located in the atop site. Finally, by a background pressure

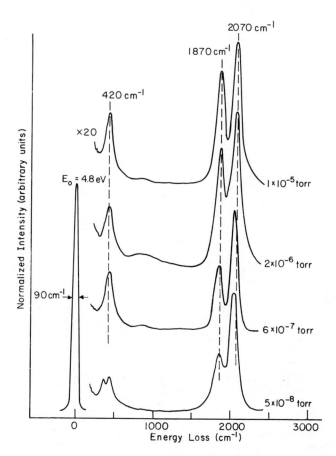

Figure 3.   Vibrational spectra of CO chemisorbed on Rh(111) at 300 K as a function of background gas pressure. The loss above 2000 cm^{-1} reaches a limiting value of 2070 cm^{-1}, while the peak at 1870 cm^{-1} increases in intensity at a constant frequency.

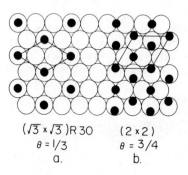

Figure 4.   Real space representations of CO chemisorbed on a Rh(111) surface: (a) $(\sqrt{3} \times \sqrt{3})R30°$ overlayer structure visible at low CO exposures; (b) (2×2) structure seen at high surface coverage

of approximately 1 x 10^{-6} torr CO at 300 K, a (2x2) LEED pattern
forms whose unit cell consists of three carbon monoxide molecules--
two atop and one bridged (Figure 4b), in reasonable agreement
with the two-to-one peak intensity ratio found in the ELS spectra
(Figure 3).

   CO Chemisorption on Pretreated Rh(111) (23).   Sexton
and Somorjai showed that surface pretreatment had a marked effect
on the rate of hydrocarbon formation from H_2/CO mixtures over
polycrystalline rhodium foils (36): oxidation enhanced the
methanation rate while surface carbon inhibited product formation.
We studied the effects of hydrogen, oxygen and carbon on the CO
on Rh(111) vibrational spectra in the hope of understanding the
effects of pre-adsorption. An extensive discussion of these
findings is presented in reference 23 and we only summarize our
results here. H_2 pre-adsorption or post-adsorption on Rh(111)
at 300 K had no significant effect on the CO vibrational spectra.
Furthermore, no room temperature rhodium-hydrogen stretching
vibrations were observed, even at H_2 exposures up to several
thousand Langmuirs. Finally, no changes were seen after heating
the crystal to 600 K in 1 x 10^{-5} torr of a 3:1 H_2/CO mixture for
30 minutes (61).
   Oxygen had a significant effect on the carbon monoxide
vibrational spectrum, however. O_2 chemisorption on Rh(111) is
dissociative at 300 K yielding a single metal-oxygen stretching
vibration at 520 cm^{-1} (23) and a second order thermal desorption
maximum (26, 62). The formation of bridge bonded carbon monoxide
was strongly inhibited in the presence of chemisorbed oxygen and
the atop sites seemed to saturate with CO by an exposure of only
1 L. It appeared that pre-adsorbed oxygen blocked some of the
surface sites so that CO could not adsorb in many of the atop and
bridged positions. Since oxygen is strongly electron withdrawing,
the extent of rhodium-carbon backbonding decreased and the $C\equiv O$
stretch shifted approximately 50 cm^{-1} to higher frequency. The
strength of the metal-adsorbate bond is determined by the electron
density in both the 5σ and $2\pi^*$ molecular orbitals of carbon
monoxide and should therefore decrease as well. Consistent with
this was a decrease of at least 30 cm^{-1} in the frequency of the
metal-carbon stretching vibration and a lowering of the CO thermal
desorption temperature by approximately 40 K (23). We found a
smaller CO thermal desorption peak area (23) and this is in
agreement with fewer CO molecules on the surface in the presence
of chemisorbed oxygen.
   The Rh(111) surface was covered with carbon by decomposing
5 x 10^{-7} torr of either acetylene or ethylene at 1100 K for 10
minutes and subsequent flashing to 1200 K (23). Pre-adsorbed
carbon had a very strong inhibiting effect on carbon monoxide
chemisorption. This is the same effect it had on the methanation
rate (36). The low inelastic scattering intensity indicated
relatively small CO coverages while the broad elastic peak and

high background level were indicative of poor ordering. Consistent with this was a high background intensity in the LEED pattern
and a decrease in the CO thermal desorption peak area. The carbon
overlayer covered most of the crystal face so that there were only
a few sites open for CO chemisorption. There was also an electronic interaction between the carbon overlayer and the adsorbed
carbon monoxide molecules, since the vibrational peaks shifted
slightly and the maximum CO thermal desorption temperature dropped
about 10 K (23).

Thus we may conclude that the pre-adsorption of hydrogen had
no effect on CO chemisorption on rhodium while both oxygen and
carbon blocked many sites for CO chemisorption and weakened the
metal-adsorbate interaction (ν_{Rh-C} decreased, $\nu_{C\equiv O}$ increased,
TDS peak maximum shifted to lower temperature).

Correlation Between Structure and C $\equiv$ O Vibrational
Frequency. The accepted picture of carbon monoxide bonding to
metals is by electron transfer from the 5σ orbital of CO to the
metallic d orbitals and by backbonding of the metallic electrons
into the empty $2\pi^*$ orbital of the adsorbate (38). This scheme
has been used by both surface scientists and inorganic chemists
to explain the infrared spectra of chemisorbed carbon monoxide
and of metal carbonyls. Since the electron density in the CO
antibonding orbital is increased, the carbon-oxygen stretching
frequency should decrease below the gas phase value of 2143 cm^{-1}.
Furthermore, as the CO is bound to an increasing number of metal
atoms this frequency should drop even further as shown in the
IR spectra of model organometallic compounds of known molecular
structure (38).

It is generally assumed that species with C-O stretching
frequencies above 2000 cm^{-1} correspond to linearly bonded CO,
frequencies between about 1850 and 2000 cm^{-1} belong to bridge
bonded species and those below approximately 1850 cm^{-1} are for
face bridging or three-fold coordination (38). Although, the
validity of this rule has not been well tested on single crystal
surfaces, some preliminary data to support these divisions has
been obtained (see Table I). This correlation of the structure
of carbon monoxide (determined by low energy electron diffraction) with the C $\equiv$ O stretching frequency (from both IR and high
resolution ELS studies) is very limited at present. Clearly more
work needs to be done before any simple interpretation of carbon-
oxygen vibrational frequencies can be used to infer the structure
of chemisorbed CO. One must also recall that the presence of
other either electron donating or withdrawing substituents on
the surface can alter the CO electron density and will certainly
shift the observed stretching frequencies (23, 69, 70).

Finally, LEED, TDS and UPS studies on the interaction of
carbon monoxide with the hexagonally closest packed faces of
the group VIII metals show numerous similarities. This is not

true of the vibrational spectroscopic data.   These results are
sumarized in Table II.   CO almost always forms a $(\sqrt{3}x\sqrt{3})R30°$

**Table I.   Correlation of Structure and Vibrational Frequency for CO
Chemisorbed on Transition Metal Surfaces**

Surface	Overlayer Structure	Position from LEED	Vibrational Frequency (cm^{-1})
Ni(100)	c(2x2)	atop ([63], [64])	2069 ([17])
Cu(100)	c(2x2)	atop ([64])	2079 ([65]), 2089 ([66])
Rh(111)	$(\sqrt{3}x\sqrt{3})R30°$	atop ([60])	2020 ([23])
Pd(100)	c(4x2)R45°	bridge ([67])	1949 ([68]), 1903 ([67])

surface structure at low coverages ([26], [59], [71-79]).   This LEED
pattern compresses through a number of intermediate steps into a
hexagonal closest packed overlayer of carbon monoxide molecules.
This is the case despite varying electronic configurations and
different metal—metal distances ([80]).   The metal—adsorbate bond
energies derived from TDS measurements vary by only ±3 kcal/mole
on the surfaces where no CO decomposition is detected ([26], [59],
[73], [76], [78], [79], [81]).   Furthermore, the binding energy difference
between the 4σ and 5σ carbon monoxide molecular orbitals, $\Delta(4\sigma-5\sigma)$, varies by only ±0.3 eV ([77], [82-88]).   The vibrational spectra
show tremendous differences, however.   Both nickel ([89]) and palla-
dium ([68]) form multiply coordinated carbonyl species at low CO
exposures and the atop species are only seen at high coverage.
The CO chemisorption behavior on Rh(111) ([23]) and Pt(111) ([90],
[91]) are the opposite: here the atop sites populate first and
predominate at low CO exposures.   Bridge bonded species begin to
form at intermediate coverages.   Ruthenium is totally different:
only a single carbon—oxygen stretching vibration is present at
all coverages ([11], [92]).   The reasons for these differences in the
nature of CO bonding to the various transition metal surfaces
will have to be explored further in the future.

The Chemisorption of Acetylene and Ethylene on Rh(111) ([24])

   Introduction.   The structure of acetylene and ethylene
adsorbed on transition metal surfaces is of fundamental importance
in catalysis.   An understanding of the interaction of these simple
molecules with metal surfaces may provide information on possible
surface intermediates in the catalytic hydrogenation/dehydrogena-
tion of ethylene.   High resolution ELS is a particularly useful

Table II. Chemisorption of CO on the Group VIII Metals

Surface	Nearest neighbor distance (Å) (80)	LEED	E_d (kcal/mole)	$\Delta(4\sigma-5\sigma)$	Vibrational Spectra (cm^{-1}) $\nu_{M\equiv C}$	ν_{CO}
Fe(111)	2.48	(1x1) (71) decomposed	24 (71)	3.2 (82)		
Ru(0001)	2.65	($\sqrt{3}$ x$\sqrt{3}$)R30° (72, 73) (2$\sqrt{3}$x2$\sqrt{3}$)R30° (5$\sqrt{3}$x5$\sqrt{3}$)R30°	28 (73)	3.1 (83, 84)	445	1984 (92) 2080
Co(0001)	2.50	($\sqrt{3}$x$\sqrt{3}$)R30° (74) hexagonal	25 (74)			
Rh(111)	2.69	($\sqrt{3}$x$\sqrt{3}$)R30° (26, 59) split (2x2) (2x2)	31 (56, 59)	3.2 (85)	480 420	1990 2070 1870 (23)
Ir(111)	2.71	($\sqrt{3}$x$\sqrt{3}$)R30° (75) (2$\sqrt{3}$x2$\sqrt{3}$)R30° split (2$\sqrt{3}$x2$\sqrt{3}$)	29.5 (81)	2.7 (86)		
Ni(111)	2.49	($\sqrt{3}$x$\sqrt{3}$)R30° (76, 77) c(4x2) ($\sqrt{7}$x$\sqrt{7}$)R19.2°	26 (76)	2.8 (77)	400	1810 1910 2050 (89)
Pd(111)	2.75	($\sqrt{3}$x$\sqrt{3}$)R30° (78) c(4x2) split (2x2) (2x2)	30.1 (78)	3.3 (87)		1823 1946 2092 (68)
Pt(111)	2.77	($\sqrt{3}$x$\sqrt{3}$)R30° (79) c(4x2) hexagonal	28 (79)	3.0 (88)	476 380	2089 1856 (90, 91) 1872

probe for studying the hydrocarbon–metal interaction because of
both its sensitivity to hydrogen and its broad spectral range
(which includes M–C stretching vibrations, C–C stretching
vibrations and C–H stretching and bending vibrations). Here we
demonstrate the power of this technique by reviewing the results
of a detailed investigation on the chemisorption and reactivity
of C_2H_2 and C_2H_4 on the Rh(111) single crystal surface (24).

LEED Studies of Acetylene and Ethylene Adsorption on Rh(111).

Exposing the clean Rh(111) surface between 230 and 250 K
to either C_2H_2 or C_2H_4 results in the appearance of sharp half
order diffraction spots in the LEED pattern from a (2x2) surface
structure. The new diffraction spots from the ordered hydrocarbon
structures are sensitive to surface coverage. Although the spots
are visible after a 1 L gas exposure, they do not become sharp
and intense until 1.5 L and then immediately begin disordering
above 1.5 L.

A diffraction pattern corresponding to a c(4x2) surface
structure can be generated from the (2x2) surface structure
without additional hydrocarbon exposure. For adsorbed C_2H_4 the
transformation occurs in vacuum by slowly warming the crystal to
300 K over the course of several hours. Rapid heating results
in the formation of a disorderd c(4x2) structure (broad, diffuse
diffraction features and some streaking). For adsorbed C_2H_2 even
this slow warm-up results in a disordered overlayer. To form a
well ordered c(4x2) structure from adsorbed acetylene
is annealed for ~4 minutes at 273 K in 1 x 10^{-8} torr of H_2 with
the mass spectrometer filaments on. These filaments are located
approximately 5 cm. from the crystal and provide the surface with
a good source of atomic hydrogen (93).

In the transformation from the (2x2) to the c(4x2) structures
the orientation and shape of the unit cell change (94), but the
areas of the primitive unit cells of these two structures are
the same (25 $Å^2$ on Rh(111)). Thus, no variation in the surface
coverage occurs. Furthermore, AES shows that the carbon coverage
from the (2x2) structures produced during C_2H_2 and C_2H_4 chemisorption are the same and remain constant during the conversion
to the c(4x2) structures (24). Thus, changes in binding site and
adsorbate geometry are probably taking place without any change
in the adsorbate coverage. The transformation from the (2x2)
to the c(4x2) surface structures is irreversible; once the
c(4x2) structure forms, the crystal can be cooled to 210 K with
no visible changes in the diffraction pattern. The c(4x2)
structures can only be altered by heating the crystal above 420 K
which causes the surface to irreversibly disorder.

ELS Studies of Acetylene Chemisorption on Rh(111).

The
vibrational spectrum of the (2x2) hydrocarbon surface structure
formed from the chemisorption of C_2H_2 on Rh(111) between 210
and 270 K is shown in Figure 5a. The peak positions and their

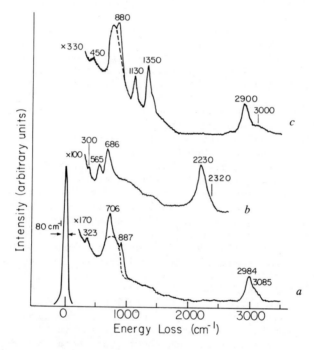

Figure 5.   High resolution ELS spectra of chemisorbed acetylene on Rh(111): (a) (2 × 2)–C_2H_2; (b) (2 × 2)–C_2D_2; and (c) c(4 × 2)–C_2H_2 + H. The vibrational assignments for Spectra (a) and (b) are listed in Table III. Spectrum (c) is discussed in more detail in the next section.

relative intensities (95) are listed in Table III. Although some
of the peaks are not readily visible in this figure, their positions and intensities are obtained from the analysis of at least

**Table III. Vibrational Modes for the Metastable (2×2) C_2H_2 (C_2D_2)
Species Observed on Rh(III) at $T \leq 270$ K (all frequencies in cm^{-1})**

C_2H_2 (C_2D_2)/Rh(111)	Assignments
3085 (~2320) w	
	C–H (C–D) stretch
2984 (2230) m	
a.	C–C stretch
887 (686) m	
	C–H (C–D) bend
706 (565) m	
323 (~300) w	M–C stretch

Intensity: s = strong   m = medium   w = weak

a.   Small broad peak in the 1300–1400 cm^{-1} region is
observed in several spectra (24).

six spectra. A complete analysis of the low frequency region in
this spectrum is hampered by a "spurious background peak" near
800 cm^{-1}. This apparent loss, first observed by Froitzheim
et al., may be caused by electon reflection from the outer half
of the analyzer (25). The dashed lines in Figures 5 through 8
indicate the approximate location and magnitude of this peak.
As a result of this experimental artifact, both the position and
intensity of all loss features between 650 and 900 cm^{-1} are rather
uncertain. Isotopic substitution is of some help in assigning the
observed vibrational frequencies to normal modes of the adsorbed
species. The ELS spectrum of the (2x2) C_2D_2 surface structure
is shown in Figure 5b.
    The vibrational spectra of the (2x2) acetylene overlayer in
Figures 5a and 5b do not change up to 270 K in vacuum. Furthermore, the positions and relative intensities of the observed
energy loss peaks are independent of the acetylene exposure (from
>0.2 L to <50 L) and, more importantly, are independent of surface
order (as determined by observation of the LEED pattern). Thus,
using high resolution ELS we conclude that the bonding of the
adsorbed molecules do not change upon disordering.

The frequencies of the carbon-hydrogen (carbon-deuterium) stretching vibrations can be used to characterize the state of hybridization of the adsorbed species. Acetylene, C_2H_2 (C_2D_2), is sp hybridized in the gas phase and has C-H (C-D) stretching vibrations between 3289 and 3374 (2439 and 2701) cm^{-1} [98]; ethylene, C_2H_4 (C_2D_4), is sp^2 hybridized and has C-H (C-D) stretching vibrations between 2989 and 3106 (2200 and 2345) cm^{-1} [98]; while ethane, C_2H_6 (C_2D_6), is sp^3 hybridized and has C-H (C-D) stretching vibrations between 2896 and 2985 (2083 and 2235) cm^{-1} [98]. Thus, the losses at 2980 (2230) and 3085 (~2320) cm^{-1} in Figure 5a and 5b and Table III correspond to the C-H (C-D) stretching vibrations of a molecule near sp^2 hybridization. This indicates that the $C \equiv C-H$ ($C \equiv C-D$) bond in adsorbed acetylene is no longer linear. The low frequency mode at 323 cm^{-1} does not shift significantly upon deuteration (~20 cm^{-1}) and most likely corresponds to the entire molecule vibrating against the surface. The two largest peaks in the spectrum at 706 and 887 cm^{-1} shift by almost 200 cm^{-1} (to 565 and 686 cm^{-1}, respectively) when C_2D_2 is chemisorbed and can be assigned to C-H (C-D) bending modes. A more detailed discussion of these mode assignments including reference to the IR spectra of model organometallic compounds is presented in reference 24. We assume the adorbate is oriented with its carbon-carbon axis approximately parallel to the surface since only small, broad peaks (1300 - 1400 cm^{-1}) are seen in the C-C stretching region. Observation of such a mode in the specular direction is prohibited by the normal dipole selection rule (3, 4) if the $C \equiv C$ bond is parallel to the surface. The vibrational mode assignments are summarized in Table III.

Bond lengths, bond angles and the position of adsorbed C_2H_2 on the surface cannot be accurately determined without a complete dynamical LEED intensity analysis. Nevertheless, the high resolution ELS results indicate that acetylene chemisorbs on Rh(111) below 270 K with its $C \equiv C$ axis oriented approximately parallel to the surface. The molecule is near sp^2 hybridization and therefore the C≡C-H bond angle is no longer linear. A similar C_2H_2 geometry is seen in numerous organometallic cluster compounds (33, 34).

Both LEED and ELS indicate that the (2x2) acetylene overlayer is stable on the surface in vacuum between 210 and 270 K. The addition of H_2 to adsorbed C_2H_2 below ~260 K causes no changes in the observed ELS spectra, although this surface species is still quite reactive. The addition of H_2 to chemisorbed C_2D_2 below 260 K results in a complex vibrational spectrum with peaks in both the C-H and C-D stretching and bending regions. Although the deuterium and hydrogen readily exchange, no change in the adsorbate geometry is detected by high resolution ELS. The vibrational spectra of adsorbed acetylene only begin to change when the crystal is heated above 270 K in vacuum. The (2x2)-C_2H_2 surface structure also disorders at this temperature.

The vibrational spectrum from the c(4x2) acetylene overlayer is shown in Figure 5c. This spectrum can either be obtained by warming the (2x2) acetylene overlayer to ~270 K in the presence of 1 x 10^{-8} torr of hydrogen or by chemisorbing C_2H_2 on Rh(111) above 300 K. Hydrogen addition to the surface species above 270 K is necessary to obtain good quality, intense ELS spectra, however. Hydrogen addition was also required to complete this conversion in the LEED studies. This species is stable on the surface up to ~240 K. The structure of this hydrocarbon overlayer will be discussed in more detail in the next section.

ELS Studies of Ethylene Chemisorption on Rh(111). The vibrational spectra from the (2x2) and c(4x2) ethylene surface structures are shown in Figures 6a and 6b. The ELS spectrum in Figure 6a is obtained by chemisorbing C_2H_4 on the crystal below 270 K. Spectrum 6b can either be observed by slowly warming the (2x2) overlayer structure (6a) to room temperature or by simply adsorbing ethylene on the Rh(111) surface above 290 K. Small peaks in the 1800 to 2100 cm^{-1} region are due to background CO adsorption (23). Once again the observed vibrational frequencies are independent of surface order and hydrocarbon exposure (<0.2 to >50 L). Note that these ELS spectra are almost identical to the vibrational spectrum from the stable c(4x2) acetylene overlayer shown in Figure 5c. The hydrocarbon species derived from ethylene chemisorption is also stable on the surface up to ~420 K. Degradation of both the c(4x2) LEED pattern and of the vibrational spectrum occur at this temperature.

The ELS spectrum resulting from the chemisorption of either C_2H_4 or C_2H_2 and H_2 on Pt(111) above room temperature are quite similar (93, 99). This species is similar to the hydrocarbons species obtained from the adsorption of ethylene on Rh(111). This is clearly shown in Figure 7. Although the chemisorption of ethylene on Pt(111) has been studied by numerous techniques (93, 99, 100, 101, 102, 103), there is still debate over the precise geometry of the stable surface species. Proposed structures include ethylidyne ($\rightarrow$C–CH_3) (100, 101), ethylidene (>CH–CH_3) (93, 99), and a vinyl species (>CH–CH_2–) (102, 103). We simply point out that the stable hydrocarbon overlayer formed from the chemisorption of either ethylene or acetylene and hydrogen on both Pt(111) and Rh(111) yield identical vibrational spectra (Figure 7). A more complete discussion of the similarities between the chemisorption of ethylene on Rh(111) and Pt(111) is presented elsewhere (24).

It is interesting to note that the geometry of the adsorbed ethylene species on Rh(111) remains the same (as indicated by the ELS spectra) while the overlayer structure changes from a (2x2) to a c(4x2). Although this conversion is not affected by the presence of hydrogen, H-D exchange will occur in the hydrocarbon overlayer when H_2 is added to chemisorbed C_2D_4. No change in the adsorbate geometry is detected by high resolution ELS.

The stable ethylene or acetylene plus hydrogen overlayer on Rh(111) can be decomposed to surface CH (CD) species above ~420 K.

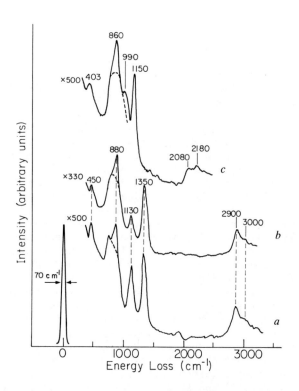

Figure 6.   High resolution ELS spectra of chemisorbed ethylene on Rh(111): (a)
(2 × 2) from C_2H_4 chemisorption; (b) c(4 × 2) from C_2H_4 chemisorption; and (c)
(2 × 2) from C_2D_4 chemisorption

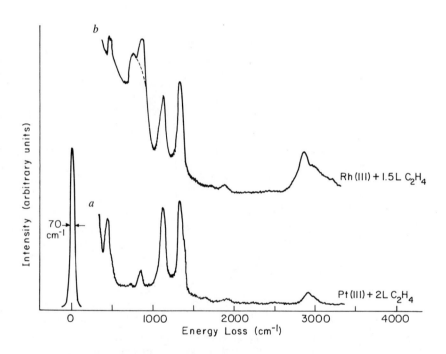

Figure 7. Comparison of the vibrational spectra for ethylene chemisorbed on: (a) Pt(111) (93); and (b) Rh(111) (24). A discussion of the similarities between acetylene and ethylene chemisorption on Rh(111) and Pt(111) is presented in Ref. 24.

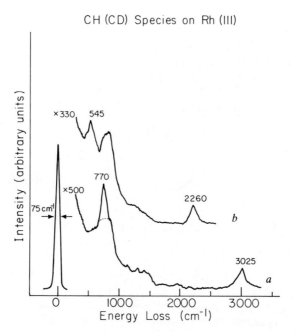

Figure 8.   CH (CD) species can be formed on Rh(111) by heating chemisorbed (a) C_2H_2 or (b) C_2D_4 to 450 K. The peak assignments are discussed in Ref. 104.

The ELS spectra for these two hydrocarbon fragments are shown in Figure 8. Assignment of the observed vibrational frequencies is discussed in detail by Demuth and Ibach for the decomposition of acetylene on Ni(111) (104). It is possible that species such as these are important surface intermediates under high pressure catalytic conditions (105, 106). Further studies in this area are in progress.

Summary. These investigations lead to the following conclusions:

1. The chemisorption of acetylene and ethylene on Rh(111) yields a series of ordered structures:

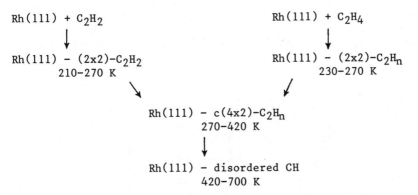

where C_2H_n stands for the stable hydrocarbon species with undetermined hydrogen content. These order-order transformations are irreversible.

2. Below 270 K acetylene chemisorbs on Rh(111) with its $C \equiv C$ bond oriented parallel to the surface forming an approximately sp^2 hybridized species.

3. Adsorption of C_2H_2 + H or C_2H_4 above 300 K on Rh(111) produces the same stable species. TDS studies show this species to have a stoichiometry of C_2H_n, where n is less than 4 (24). Adsorption of C_2H_2 + H or C_2H_4 on Pt(111) above 350 K leads to a similar, and most likely identical, species.

4. The geometry of the adsorbed ethylene species on Rh(111), as determined by ELS, does not change during the conversion from the metastable to the stable species although the overlayer structure changes from a (2x2) to a c(4x2).

5. The addition of H_2 to chemisorbed C_2D_2 or C_2D_4 results in H-D exchange, but no change in the adsorbate geometry is detected by ELS.

Conclusion

In this paper we have presented several applications of high resolution electron energy loss spectroscopy to the characteriza-

tion of adsorbed molecules on single crystal surfaces. ELS is
clearly a very powerful technique for studying the vibrational
spectra of chemisorbed molecules. However, no single surface
sensitive probe alone can yield a complete picture of the
structure, bonding and reactivity of adsorbed species. Experiments utilizing a combination of complementary surface sensitive
techniques, in particular ELS and LEED, will yield more detailed
structural information about adsorbed atoms and molecules
in the future.

Acknowledgments

We thank Prof. P. K. Hansma and Dr. D. G. Castner for many
enlightening discussions and for valuable technical assistance.
This work was supported by the Division of Materials Science,
Office of Basic Energy Sciences, U.S. Department of Energy.

Literature Cited

1.   Ertl, G.; Kuppers, J. "Low Energy Electrons and Surface
     Chemistry," Verlag Chemie: Germany, 1974.
2.   Ibach, H., Ed. "Electron Spectroscopy for Surface Analysis,"
     Springer-Verlag: Berlin, 1977.
3.   Evans, E.; Mills, D. L. Phys. Rev., 1972, B5, 4126.
4.   Ibach, H. Surface Sci., 1977, 66, 56.
5.   Sokcevic, D.; Lenac, Z.; Brako, R.; Sunjic, M. Z. Physik.,
     1977, B28, 273.
6.   Sunjic, M.; Brako, R., Lenac, Z.; Sokcevic, D. Intern. J.
     Quantum Chem., 1977, Suppl. 2, 59.
7.   Lenac, Z.; Sunjic, M.; Sokcevic, D.; Brako, R. Surface
     Sci., 1979, 80, 602.
8.   Davenport, J. W.; Ho, W.; Schrieffer, J. R. Phys. Rev.,
     1978, B17, 3115.
9.   Ho, W.; Willis, R. F; Plummer, E. W. Phys. Rev. Letters,
     1978, 40, 1463.
10. Willis, R. F.; Ho, W.; Plummer, E. W. Surface Sci., 1979,
     88, 384.
11. See chapter by P. A. Thiel and W. H. Weinberg and references
     therein.
12. Propst, F. M.; Piper, T. C. J. Vac. Sci. Technol., 1967,
     4, 53.
13. Ibach, H. "Proceedings of the Conference on Vibrations
     in Adsorbed Layers," Julich, Germany, 1978; p. 64 and
     references therein.
14. Dalmai-Imelik, G.; Bertolini, J. C.; Rousseau, J. Surface
     Sci., 1977, 63, 67.
15. Backx, C.; Feuerbacher, B.; Fitton, B.; Willis, R. F.
     Surface Sci., 1977, 63, 193.
16. Thomas, G. E.; Weinberg, W. H. Rev. Sci. Instrum., 1979,
     50, 497.

17. Andersson, S. Solid State Commun., 1977, 21, 75.
18. Andersson, S. Surface Sci., 1979, 79, 385.
19. Roy, D.; Carette, J. D. in Ibach, H., Ed. "Electron Spectroscopy for Surface Analysis," Springer-Verlag: Berlin, 1977; p. 13.
20. Froitzheim, H.; Ibach, H.; Mills, D. L. Phys. Rev., 1975, B11, 4980.
21. Adnot, A. "Proceedings of the Conference on Vibrations in Adsorbed Layers," Julich, Germany, 1978; p. 109.
22. Dubois, L. H.; Hansma, P. K.; Somorjai, G. A., to be published.
23. Dubois, L. H.; Somorjai, G. A. Surface Sci., 1980, 91, 514.
24. Dubois, L. H.; Castner, D. G.; Somorjai, G. A. J. Chem. Phys., 1980, 72, 000.
25. Froitzheim, H.; Ibach, H.; Lehwald, S. Rev. Sci. Instrum. 1975, 46, 1325.
26. Castner, D. G.; Sexton, B. A.; Somorjai, G. A. Surface Sci. 1978, 71, 519.
27. Ford, R. R. Advan. Catalysis, 1970, 51, 21.
28. Bradshaw, A. M., Surface Sci., 1979, 80, 215.
29. King, R. B. Prog. Inorg. Chem., 1972, 15, 287.
30. Chini, P.; Longini, G.; Albano, V. G. Adv. Organomet. Chem., 1976, 14, 285.
31. Ozin, G. A. Acc. Chem. Res., 1977, 10, 21.
32. Muetterties, E. L. Bull. Soc. Chim. Belg., 1975, 84, 959.
33. Muetterties, E. L. Bull. Soc. Chim. Belg., 1976, 85, 451.
34. Muetterties, E. L.; Rhodin, T. N.; Band, E.; Brucker, C. F.; Pretzer, W. R. Chem. Rev., 1979, 79, 91.
35. Bhasin, M. M.; Bartley, W. J.; Ellgen, P. C.; Wilson, T. P. J. Catal., 1978, 54, 120.
36. Sexton, B. A.; Somorjai, G. A. J. Catal., 1977, 46, 167.
37. Pruett, R. L. Ann. N. Y. Acad. Sci., 1977, 295, 239.
38. See Sheppard, N.; Nguyen, T. T. in Clark, R. J. H.; Hester, R. E., Eds. "Advances in Infrared and Raman Spectroscopy" Vol. 5, Heydon and Son: London, 1978; p. 67 and references therein.
39. Yang, A. C.; Garland, C. W. J. Phys. Chem. 1957, 61, 1504.
40. Arai, H.; Tominaga, H. J. Catal., 1976, 43, 131.
41. Yao, H. C.; Rothschild, W. G. J. Chem. Phys., 1978, 68, 4774.
42. Primet, M. J. C. S. Faraday I, 1978, 74, 2570.
43. Yates, J. T., Jr.; Duncan, T. M.; Worley, S. D.; Vaughn, R. W. J. Chem. Phys., 1979, 70, 1219.
44. Yates, D. J. C.; Murrell, L. L.; Prestridge, E. B. J. Catal., 1979, 57, 41.
45. Smith, G. C.; Chojnacki, T. P.; Drasgupta, S. R.; Iwatata, K.; Watters, W. L. Inorg. Chem., 1975, 14, 1419.
46. Conrad, H.; Ertl, G.; Knozing, H.; Kuppers, J.; Latta, E. E. Chem. Phys. Lett., 1976, 42, 115.

47. Hansma, P. K.; Kaska, W. C.; Laine, R. M. J. Amer. Chem. Soc., 1976, 98, 6064.
48. Klein, J.; Leger, A.; de Cheveigne, S.; Guinet, C.; Belin, M.; Defourneau, D. Surface Sci., 1979, 82, L288.
49. Kroeker, R. M.; Kaska, W. C.; Hansma, P. K.; J. Catal., 1979, 57, 72.
50. Whyman, R. J. C. S. Chem. Comm., 1970, 1194.
51. Griffith, W. P.; Wickam, A. J. J. Chem. Soc. A, 1969, 834.
52. Hanlan, L. A.; Ozin, G. A. J. Amer. Chem. Soc., 1974, 96, 6324.
53. Roth, J. G.; Dignam, M. J. Can. J. Chem., 1976, 54, 1388.
54. Moskovitz, M.; Hulse, J. W. Surface Sci., 1978, 78, 397.
55. Scheffler, M. Surface Sci., 1978, 78, 397.
56. Blyholder, G. J. Phys. Chem., 1964, 68, 2772.
57. Crossley, A.; King, D. A. Surface Sci., 1977, 68, 528.
58. Hollins, P.; Pritchard, J. Surface Sci., 1979, 89, 486.
59. Thiel, P. A.; Williams, E. D.; Yates, J. T., Jr.; Weinberg, W. H. Surface Sci., 1979, 84, 54.
60. Van Hove, M. A.; Koestner, R. J.; Somorjai, G. A., to be published.
61. The Rh(111) surface is catalytically active for Fischer-Tropsch synthesis, but at pressures above one atmosphere (Castner, D. G.; Blackadar, R. L.; Somorjai, G. A., submitted to J. Catal.).
62. Thiel, P. A.; Yates, J. T., Jr.; Weinberg, W. H. Surface Sci., 1979, 82, 22.
63. Passler, M.; Ignatiev, A.; Jona, F.; Jepsen, D. W.; Marcus, P. M. Phys. Rev. Lett., 1977, 43, 360.
64. Andersson, S.; Pendry, J. B. Phys. Rev. Lett., 1977, 43, 363.
65. Horn, K.; Pritchard, J. Surface Sci., 1976, 55, 701.
66. Sexton, B. A., Chem. Phys. Lett., 1979, 63, 451.
67. Behm, R. J.; Christmann, K.; Ertl, G.; Van Hove, M. A.; Thiel, P. A.; Weinberg, W. H. Surface Sci., 1979, 88, L59.
68. Bradshaw, A. M.; Hoffman, F. M. Surface Sci. 1978, 72, 513.
69. Wojtczak, J.; Queau, R.; Poilblane, R. J. Catal., 1975, 37, 391.
70. Ibach, H.; Somorjai, G. A. Appl. Surface Sci., 1979, 3, 293.
71. Yoshida, K.; Somorjai, G. A. Surface Sci., 1978, 75, 46.
72. Williams, E. D.; Weinberg, W. H. Surface Sci., 1979, 82, 93.
73. Madey, T. E.; Menzel, D. Japan J. Appl. Phys. Supp. 2, Part 2, 1974, 229.
74. Bridge, M. E.; Comrie, C. M.; Lambert, R. M.; Surface Sci., 1977, 67, 393.
75. Kuppers, J.; Plagge, A. J. Vac. Sci. Technol., 1976, 13, 259.

76. Christmann, K.; Schober, O.; Ertl, G. J. Chem. Phys., 1974, 60, 4719.
77. Conrad, H.; Ertl, G.; Kuppers, J.; Latta, E. E. Surface Sci., 1976, 57, 475.
78. Conrad, H.; Ertl, G.; Koch, J.; Latta, E. E. Surface Sci., 1974, 43, 462.
79. Ertl, G.; Neumann, M.; Streitt, K. M. Surface Sci, 1977, 64, 393.
80. Kittel, C. "Introduction to Solid State Physics," Wiley, New York, 1976; p. 32.
81. Nieuwenhuys, B. E.; Hagen, D. I.; Rovida, G.; Somorjai, G. A. Surface Sci. 1976, 59, 155.
82. Textor, H. M.; Gay, I. D.; Mason, R. Proc. Royal Soc. London, Ser. A., 1977, 356, 37.
83. Fuggle, J. C.; Madey, T. E.; Steinkilberg, M.; Menzel, D. Phys. Lett. 1975, 51A, 163.
84. Fuggle, J. C.; Madey, T. E.; Steinkilberg, M.; Menzel, D. Surface Sci., 1975, 52, 521.
85. Braun, W.; Neumann, M.; Iwan, M.; Koch, E. E. Solid State Commun., 1978, 27, 155.
86. Zhdan, P. A.; Boreskov, G. K.; Baronin, A. I.; Egelhoff, W. F., Jr.; Weinberg, W. H. Chem. Phys. Lett., 1976, 44, 528.
87. Lloyd, D. R.; Quinn, C. M.; Richardson, N. V. Solid State Commun. 1976, 20, 409.
88. Iwasawa, Y.; Mason, R.; Textor, M.; Somorjai, G. A. Chem. Phys. Lett., 1976, 44, 468.
89. Erley, W.; Wagner, H.; Ibach, H. Surface Sci., 1979, 80, 612.
90. Froitzheim, H.; Hopster, H.; Ibach, H.; Lehwald, S. Appl. Phys. 1977, 13, 147.
91. Hopster, H.; Ibach, H. Surface Sci., 1978, 77, 109.
92. Thomas, G. E.; Weinberg, W. H.; J. Chem. Phys., 1979, 70, 1437.
93. Ibach, H.; Lehwald, S. J. Vac. Sci. Technol., 1978, 15, 407.
94. On the (111) crystal face of an fcc metal the (2x2) unit cell is a rhombohedron while the c(4x2) primitive unit cell is a rectangle.
95. The loss intensities are a strong function of the incident electron energy and only average values are reported here. This could be due to molecular resonances or short range "impact" scattering (8, 9, 10, 96) as discussed by Lehwald and Ibach for the case of acetylene chemisorbed or Ni(111) (97).
96. Andersson, S.; Davenport, J. W. Solid State Commun., 1978, 28, 677.
97. Lehwald, S.; Ibach, H.; to be published.

98. See for example, Shimanouchi, T. "Tables of Molecular
     Vibrational Frequencies," consolidated Vol. 1, U.S.
     Department of Commerce publication NSROS-NBS 39, 1972.
99. Ibach, H.; Hopster, H.; Sexton, B. Appl. Surface Sci.,
     1977, 1, 1.
100. Kesmodel, L. L.; Dubois, L. H.; Somorjai, G. A. Chem. Phys.
     Lett. 1978, 56, 267.
101. Kesmodel, L. L.; Dubois, L. H.; Somorjai, G. A. J. Chem.
     Phys., 1979, 70, 2180.
102. Demuth, J. E. Surface Sci., 1979, 80, 367.
103. Demuth, J. E. Surface Sci., to be published.
104. Demuth, J. E.; Ibach, H. Surface Sci., 1978, 78, L238.
105. Hattori, T.; Burwell, R. L. Jr. J. Phys. Chem., 1979, 83,
     241.
106. Davis, S. M.; Somorjai, G. A., to be published.

RECEIVED June 3, 1980.

An Electron Energy Loss Spectroscopic Investigation of the Adsorption of Nitric Oxide, Carbon Monoxide, and Hydrogen on the Basal Plane of Ruthenium

PATRICIA A. THIEL and W. HENRY WEINBERG

Division of Chemistry and Chemical Engineering, California Institute of Technology, Pasadena, CA 91125

The study of the adsorption of nitric oxide on the surfaces of Group VIII transition metals is relevant to several major catalytic processes. Among these are the oxidation of ammonia to nitric oxide and subsequent synthesis of nitric acid (1,2), and the environmentally important removal of nitric oxide from exhaust stream effluents (3). The latter process involves dissociation of NO and requires the presence of a reducing agent, e.g., CO or H_2, to remove oxygen from the catalyst surface and thus prevent its being poisoned (3). Ruthenium is unique among the noble metals in its strong selectivity for formation of N_2 as a decomposition product of NO at relatively low temperatures, i.e., less than 800 K (4,5,6). This desirable property may be attributable to the ability of Ru to adsorb NO dissociatively at these temperatures (7,8), a property which has been verified by a number of spectroscopic studies of the adsorption of NO on single crystallographic surfaces of Ru (7 - 15).

The adsorption of nitric oxide is of further interest from the point of view of metal-nitrosyl interactions in inorganic chemistry. The bonding of nitrosyl ligands to transition metal centers in metal compounds has indicated that the NO ligand is amphoteric, i.e., it can be formally considered as NO^+ or NO^- (linear or bent) when bonding to a single metal center (16,17). [It should be noted that the oxidation state designation does not necessarily reflect the true charge distribution about the ligand (18)]. It has also been observed that nitrosyl ligands, like carbonyl ligands, can form bonds to more than one metal atom (16). Very recently, spectroscopic investigations have provided evidence that at least two types of molecular NO can exist simultaneously on transition metal surfaces (12,15,19,20, 21, 22), thus supporting the interesting possibility that the rich diversity of nitrosyl-metal chemistry extends from metal complexes to metal surfaces. Two states of molecular NO have been observed on Ru(001) by electron energy loss spectroscopy (EELS) (13), and also by X-ray photo-

0-8412-0585-X/80/47-137-191$06.25/0
© 1980 American Chemical Society

electron spectroscopy (XPS) and ultraviolet photoelectron spectroscopy (UPS) (16). Two apparently similar states have been observed on the Ir(111) surface using UPS and XPS (19,20), and also
on the Pt(111) and reconstructed Pt(100) surfaces using EELS (21,
22).

In this chapter, recent results are discussed in which the
adsorption of nitric oxide and its interaction with co-adsorbed
carbon monoxide, hydrogen, and its own dissociation products on
the hexagonally close-packed (001) surface of Ru have been characterized using EELS (13, 14, 15). The data are interpreted in terms
of a site-dependent model for adsorption of molecular NO at 150 K.
Competition between co-adsorbed species can be observed directly,
and this supports and clarifies the models of adsorption site
geometries proposed for the individual adsorbates. Dissociation
of one of the molecular states of NO occurs preferentially at
temperatures above 150 K, with a coverage-dependent activation
barrier. The data are discussed in terms of their relevance to
heterogeneous catalytic reduction of NO, and in terms of their
relationship to the metal-nitrosyl chemistry of metallic complexes.

Electron Energy Loss Spectroscopy

In electron energy loss spectroscopy, a highly monochromated
beam of low-energy electrons (1 - 10 eV) is reflected from a surface, and the energy of the scattered electrons is analyzed. Some
of the incident electrons (for this low range of kinetic energies)
can lose a discrete amount of energy upon reflection which can be
associated with the excitation of a vibration at the surface. The
theory based on long-range dipole scattering constituted the original framework for interpretation of inelastic electron scattering
data. In this model, if a perfect image of the dipole (adsorbate
bond) is assumed to be created in the metal, then all components
of the adsorbate dipole which are parallel to the surface cancel
exactly. Moreover, there is an image of the approaching electron
created in the bulk metal. Considering only the long-range Coul-
ombic interactions of the electron, the resulting electric field
is perpendicular to the surface near the dipole, and only those
vibrational modes of an adsorbed molecule with a component of the
dipole moment perpendicular to the surface will be excited. This
is the essence of the so-called "dipole-normal selection rule" and
is identical to that associated with reflection IR spectroscopy
(23). With this physical view of the scattering mechanism, the
cross section for scattering from a single dipole (24, 25, 26), or for
scattering from a dipole array has been calculated within the Born
approximation (27, 28, 29). The screening effect of the metal plasmons has been considered also (30). The essential result is that
the inelastically scattered electrons are peaked sharply in the
specular direction, and the relative loss intensity initially increases monotonically from zero with increasing impact energy (23 - 30).
Recent electron scattering experiments from hydrogen on W(100) (31,32),

hydrogen on Pt(111) (33), and hydroxyl groups on NiO(111) (34),
however, have shown that the observed angular dependence and
electron-impact energy dependence are not consistent with simple
dipole scattering from these surfaces. The observation of "dipole-
forbidden" vibrations in off-specular scattering (31,32,33) and of
anomalous intensities (34) have led to proposals that short-range
impact (thermal diffuse) scattering (32,33), quadrupolar scatter-
int (29), and/or resonance scattering (34) mechanisms are important. To illustrate these ideas, Fig. 1 shows the impact-energy
dependence observed by Andersson and Davenport for EEL scattering
from CO adsorbed on Ni(100) [Fig. 1(A)] and the same dependence
for scattering from OH on NiO(111) [Fig. 1(B)] (34). The predicted resonance dependence for adsorbed CO (which does not fit the
experimental data) is based on a calculation by Davenport et al.
(35) for electron scattering from oriented diatomic molecules in
the absence of a metal surface. Although a comparable calculation
is not available for OH, it is apparent in Fig. 1(B) that the
dipole theory does not provide even a qualitative approximation to
the experimentally observed fundamental and overtone intensities.
This strongly suggests that resonant scattering, or at least non-
dipolar scattering, is operative for OH on NiO(111) (34).

Experimental Methods

The results reported below were obtained in a stainless steel
ultrahigh vacuum system with a base pressure below 5×10^{-11} torr.
The Ru crystal was cut from a boule obtained from Materials Research Corporation, which was oriented to within $1°$ of the (001)
plane, as determined by Laue back-reflection X-ray diffraction, and
then cut with a rotating wire saw. It was mounted on a polishing
goniometer (36) and reoriented before polishing with standard
mechanical techniques. The crystal was immersed in boiling aqua
regia for 30 seconds before it was mounted on a precision manipulator and placed in the vacuum chamber. It was then cleaned by
heating repeatedly to 1370 K in 5×10^{-8} torr O_2, followed by
several minutes of annealing in vacuum at 1570 K. This cleaning
procedure has been suggested by previous authors (37) and was verified in our laboratory in a separate vacuum system equipped with
low-energy electron diffraction (LEED) optics and a single-pass
cylindrical mirror energy analyzer for Auger electron spectroscopy
(AES) (38,39). Occasionally, the oxygen cleaning treatment and
exposures to nitric oxide resulted in formation of a tenacious
surface oxide, as judged by the presence of a 64 to 74 meV loss
(and gain) feature in the inelastic scattering spectrum (39,40).
Heating the Ru sample for total periods of up to 30 minutes at
1580 - 1660 K was required to remove this oxide. The crystal temperature was monitored with a 95% W/5% Re vs. 74% W/26% Re thermocouple spotwelded to the edge of the Ru disk, and the thermocouple
calibration of Sandstrom and Withrow (41) was used to correlate
thermocouple voltage with temperature below 273 K. The crystal

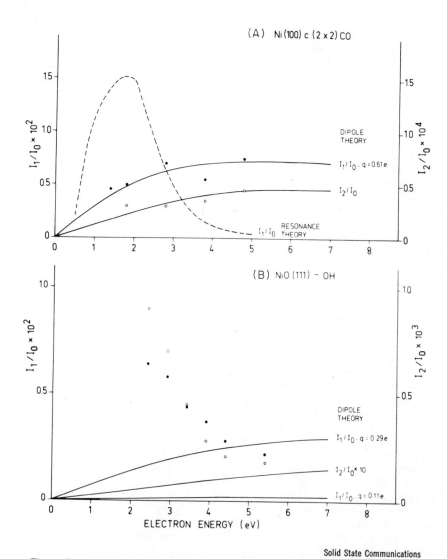

Figure 1. *(A) Relative loss intensity as a function of primary electron energy for*
Ni(100)c(2 × 2)CO derived from (●) measured fundamental and (○) overtone
loss spectra. Also shown are the (– – –) resonance and (———) dipole theories; (B)
Relative loss intensity as a function of primary electron energy for NiO(111)–OH
derived from (●) measured fundamental and (○) overtone loss spectra; (———) di-
pole predictions as explained in the text (34)

could be cooled with liquid nitrogen to approximately 150 K within
10 to 20 minutes after annealing at 1570 K. It was heated resis-
tively by passing current through the two parallel Ta support
wires, 0.025 cm in diameter, which were spotwelded on the back
face of the crystal.

The Ru sample was exposed to NO through a directional beam
doser (42). Exposures of NO, ε_{NO}, are reported in units of torr-s,
where the pressure refers to the NO pressure in a gas storage bulb
behind a glass capillary, and the time refers to the time of exposure. Coverages of NO relative to saturation, $\hat{\theta}_{NO}$, are based on
thermal desorption peak areas of N_2 and NO. Similarly, relative
coverages of hydrogen are based on integrated desorption peak
areas of H_2. Thermal desorption spectra were measured with a
quadrupole mass spectrometer, model UTI 100C, using an emission
current of 2.0 mA. Isotopic $^{15}N^{16}O$ was used in the thermal desorption experiments. The evolution of $^{15}N_2$ (mass 30) could then
be measured without interference from possible traces of CO con-
tamination (mass 28). The sample was exposed to H_2 and CO by
backfilling the entire vacuum system with the gas, and these exposures are reported in units of Langmuirs (1 L $\equiv$ 1 Langmuir $\equiv$
1 x 10^{-6} torr-s), corrected for the sensitivity, S, of the ionization gauge for H_2 and CO relative to N_2 (S_{H_2} = 0.5 and S_{CO} = 1.05)
(43). Absolute coverages of CO, θ_{CO}, have been calibrated by combining thermal desorption and LEED data (38,39).

The electron spectrometer used in these energy loss measurements, based on the Kuyatt-Simpson hemispherical deflector design
(44), has been described fully elsewhere (45). The energy of the
incident beam was 4 eV, and the angle of incidence was 62° from
the surface normal. The full-width at half-maximum of the elastic
peak varied from 10 to 16 meV, with 5 x 10^5 cps of elastic electrons reflected specularly from the clean surface. The energies
of the loss features are accurate to within approximately ±1 meV
(1 meV $\equiv$ 8.066 cm^{-1}).

Adsorption and Dissociation of Nitric Oxide (12,14)

The vibrational spectra of four different coverages of nitric
oxide chemisorbed on the Ru(001) surface are shown in Fig. 2.
(Unless otherwise specified, the temperature of adsorption in all
cases is between 120 K and 180 K.) Two loss features are observed
in the energy range in which one might expect to observe the
nitrogen-oxygen stretching mode of molecularly adsorbed NO, i.e.,
between 1350 and 2000 cm^{-1}, in analogy with the infrared spectra
of metal compounds containing nitrosyl ligands (16,17,46). The
frequencies of both these transitions increase with increasing
coverage, as shown in Fig. 3. The high-frequency vibration, which
appears at coverages greater than approximately one-third of
saturation, varies in energy from 1783 to 1823 cm^{-1}. This is in
the region usually associated with a terminal and linear M-N-O
(formally NO$^+$) configuration (16,17,46). The high-energy loss

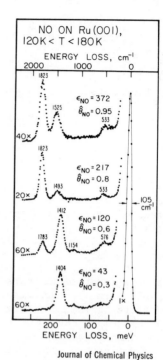

Figure 2. Electron energy loss spectra of NO adsorbed on the Ru(001) surface, for various coverages where θ_{NO} is the coverage of NO relative to saturation. The energies of individual loss features are given in cm^{-1} (13).

Journal of Chemical Physics

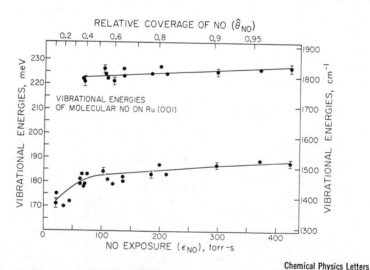

Chemical Physics Letters

Figure 3. Energies of the nitrogen–oxygen vibrations of the two major states of molecular NO on Ru(001) at 150 K as functions of exposure and coverage. The error bars show the typical uncertainty in each individual frequency assignment, approximately ±1 meV (±8 cm^{-1}). The lower curve represents the frequencies for bridged NO, and the upper curve represents the frequencies for linear NO (14).

feature can thus be reasonably assigned to molecular NO linearly bound to a single Ru atom.

The vibration at lower energies, which varies from 1379 to 1525 cm^{-1}, is due to a form of molecular NO which is populated first, at lowest coverages. In general, frequencies between 1525 and 1750 cm^{-1} are due to bent nitrosyls (formally NO^-) with M-N-O angles between 119° and 177°, where M is a transition metal atom (16,17). Although it must be kept in mind that correlations between vibrational frequencies and bonding configurations are not exact due to (at least in part) the variable electronic influence of the other ligands in the metal complex (16,17,47), it is apparent that the energy range observed for bent nitrosyl ligands (1525 - 1750 cm^{-1}) does not overlap the energy of the low-frequency vibration observed in the chemisorption experiments (1379 - 1525 cm^{-1}). The low-energy vibration of molecularly adsorbed NO may be assigned more satisfactorily to "bridged" NO, adsorbed at a site of two- or three-fold coordination. In support of this assignment, each of the nitrosyl ligands in the compound $Ru_3(CO)_{10}(NO)_2$ is bound to two metal atoms, and the N-O vibrational frequencies lie between 1533 and 1500 cm^{-1}, depending slightly upon the solvent or matrix (48, 49, 50). In the compound $(C_5H_5)_3Co_3(NO)_2$, the nitrosyls are both bound to three cobalt atoms, and a strong absorption band due to the N-O stretching frequency is found at 1405 cm^{-1} (51). The three-fold symmetry of the Ru(001) surface, combined with the infrared frequencies discussed above, suggests that the low-frequency state of molecular NO is adsorbed at a site of three-fold (or possibly two-fold) coordination. This strongly bound state will be referred to as "bridged" NO.

A transient feature at approximately 1154 cm^{-1} occurred in some spectra and apparently represents an adsorbate which is unstable between 120 K and 180 K on Ru(001). In several cobalt complexes, the presence of the hyponitrite ligand, N_2O_2, is characterized by infrared bands between 925 and 1195 cm^{-1}, depending upon the nature of the anion and other ligands (46,52,53). The exact identity of this transient species on the Ru(001) surface will require further clarification, however.

In the energy loss spectra of Fig. 2, a transition at 533 - 576 cm^{-1} is observed also. The frequency range for the Ru-O stretching mode of atomic oxygen on Ru(001), determined separately (39,40), is 516 - 596 cm^{-1}. However, in inorganic complexes the Ru-N stretching vibration of those nitrosyl ligands with vibrations above 1800 cm^{-1} is reported to occur in the range 570 - 638 cm^{-1} (47,54). In our spectra, the intensity of the low-frequency transition correlates with the appearance and growth in intensity of the loss feature above 1780 cm^{-1}, which is due to linear NO. This suggests strongly that, below 180 K, the vibrational feature at 533 - 576 cm^{-1} is due to the Ru-NO stretch of linear NO. This assignment is supported by the fact that, below 150 K, Umbach et al. (15) see no evidence for dissociation of NO on Ru(001) using X-ray photoelectron spectroscopy. At higher temperatures,

where dissociation products and molecular NO are both present on
the Ru(001) surface, the Ru-NO loss feature is unresolved from the
Ru-O feature in the EEL spectra.

In order to illustrate the steps involved in the dissociation
of NO, a series of experiments was performed in which the surface
prepared with molecularly adsorbed NO was heated briefly at approximately 5K/s and allowed to cool. The maximum temperature was
maintained for approximately one second. The surface cooled below
140 K during the recording of the vibrational spectrum, so that
only irreversible changes in the adsorbed layer could be observed.
The vibrational spectrum is shown as a function of temperature
treatment in Figs. 4 and 5 for surfaces prepared with NO coverages
(relative to saturation) of 0.3 and 0.8, respectively.

In Figs. 4(b) and 4(c), spectra are shown which illustrate the
irreversible change in the adsorbed layer as the surface is heated
to room temperature. The low-frequency N-O stretching mode disap-
pears well before complete dissociation occurs. After this surface
is heated to 407 K, dissociation is complete, as indicated by the
sole loss feature at 573 cm^{-1}. The intermediate state of the
surface [Fig. 4(c)] is characterized by a single N-O stretching
frequency at approximately 1782 cm^{-1} and a low-frequency mode at
538 cm^{-1}, depending upon coverage.

One explanation for the change in the vibrational spectrum is
that the bridged NO begins to dissociate near room temperature,
and that the N and O atoms displace the remaining NO from the
three-fold or bridged sites. The linear form of NO is less reactive and does not dissociate until higher temperatures are reached.
The metal-oxygen stretching frequency is consistent with its occupying a threefold or bridge site (39).

The mechanism of site blocking also explains results obtained
for an NO coverage of 0.8 (Fig. 5). Here, the disappearance of
the more highly coordinated NO is nearly complete only at 406 K,
after complete dissociation would have occurred for an NO coverage
of 0.3 ; and the linear NO is not removed until 500 K. The final
transformation of the saturated surface is accompanied by the
rapid evolution of NO and the beginning of N_2 desorption. The dissociation of the linear NO is prevented at high coverage by the
lack of sites for the reaction products. The transformations observed in these experiments are most likely limited kinetically in
the brief heating cycles.

Thus a relatively simple scheme may be proposed in which the
two forms of NO adsorbed on the Ru(001) surface, the relative populations of which vary with coverage, have different activation
energies for dissociation. The dissociative reaction is poisoned
by the reaction products below their temperature of desorption.

Further experiments, involving co-adsorption of molecular NO
and its own dissociation products, support this two-site model.
Energy loss spectra are shown in Figs. 6 and 7 where the Ru(001)
surface was treated sequentially as follows: (a) It was exposed to
NO at 150 K; (b) It was heated to a temperature sufficient to

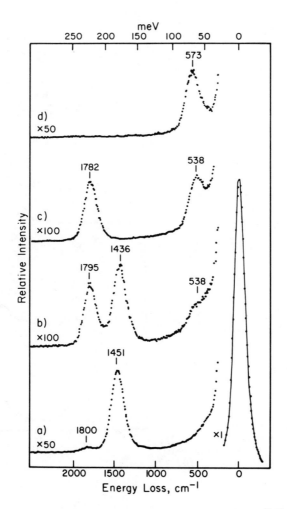

Physics Review Letters

Figure 4.   Effect of heating the surface with an initial NO coverage, θ_{NO}, of 0.3 (12): (a) after exposure at 150 K; (b) after heating to 290 K; (c) after heating to 316 K; (d) after heating to 407 K

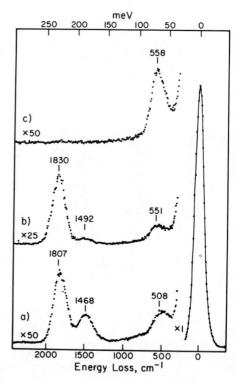

Figure 5. Effect of heating the surface with an initial NO coverage of 0.8 (12): (a) after exposure at 160 K; (b) after heating to 406 K; (c) after heating to 500 K

Physics Review Letters

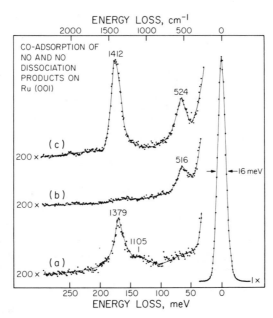

ENERGY LOSS, cm⁻¹

CO-ADSORPTION OF
NO AND NO
DISSOCIATION
PRODUCTS ON
Ru (001)

Chemical Physics Letters

Figure 6. A series of energy loss spectra of molecular NO and its dissociation products adsorbed on Ru(001) (14): (a) follows an exposure of ϵ_{NO} of 22 torr-s ($\theta_{NO} = 0.15$) at T $= 150$ K *on clean Ru(001); (b) follows heating the surface represented in (a) to 250K;(c) follows exposing the surface represented in (b) to 43 torr-s NO at 150 K*

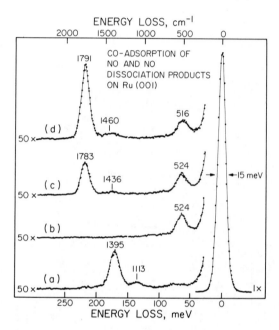

Chemical Physics Letters

Figure 7.   A series of energy loss spectra of molecular NO and its dissociation products adsorbed on Ru(001) (14): (a) follows an exposure of ϵ_{NO} of 55 torr-s ($\theta_{NO} = 0.3$) at 150 K on clean Ru(001); (b) follows heating the surface represented in (a) to 311 K; (c) follows exposing the surface represented in (b) to 43 torr-s NO at 150 K; (d) follows exposing the surface represented in (c) to 43 torr-s NO at 150 K

dissociate the adsorbed NO; and (c) The surface prepared in (b) was exposed again to NO at 150 K. The initial relative (to saturation) coverage of NO in Fig. 6(a) is $\hat{\theta}_{NO} = 0.15$. The peak at 1379 cm^{-1} is due to bridged NO. The spectrum of Fig. 6(b) follows heating the surface to 250 K. In this spectrum, the peak at 516 cm^{-1} is due to chemisorbed oxygen. When this surface was then exposed to 43 torr-s NO at 150 K, spectrum (c) of Fig. 6 resulted, with peaks at 1412 and 524 cm^{-1}. It can be seen, by comparison with Fig. 7(a), that the presence of the dissociation products resulting from the small initial NO coverage does not significantly influence the subsequent adsorption of NO, on the basis of the energy loss spectra. The dissociation products do not prevent the NO from adsorbing into the bridged state, where it would adsorb also at low coverages on the clean Ru(001) surface (Fig. 2).

At a higher initial coverage of NO, $\hat{\theta}_{NO} = 0.3$, the results are quite different. The vibrational spectrum of this initial surface coverage is shown in Fig. 7(a), with peaks at 1395 and 1113 cm^{-1}. Spectrum (b) of Fig. 7 follows heating to 311 K, which again leaves a surface with only chemisorbed nitrogen and oxygen atoms present. The spectrum of Fig. 7(c) was measured after the surface with dissociation products had been exposed to 43 torr-s of NO at 150 K. It can be seen that the higher coverage of nitrogen and oxygen adatoms resulting from the higher initial coverage of NO prevents almost completely subsequent adsorption of molecular NO into the bridged state. Rather, adsorption of linear NO occurs after this exposure, in contrast to the behavior shown in spectrum (c) of Fig. 6. Further exposure of this surface to NO leads only to a growth in the intensity of the feature at 1791 cm^{-1}, which is due to linear NO, as shown in Fig. 7(d).

The experimental data of Figs. 6 and 7 provide strong support for the two-site model of adsorption of NO. The dissociation products of NO occupy the same sites on the Ru(001) surface as the bridged state of molecular NO. When the initial coverage of dissociation products is sufficiently high, as in Fig. 7, the surface is poisoned with respect to subsequent adsorption of bridged NO. At a lower initial coverage of N and O adatoms (Fig. 6), the NO can adsorb first into the (vacant) bridged sites, which are energetically preferred. These conclusions are reminiscent of those drawn from previous UPS studies of the adsorption of NO on Ru(100) at 370 K, in which dissociation of NO was shown to precede adsorption of molecular NO in coverage (7). This adsorption sequence has also been postulated to occur on polycrystalline Rh at room temperature on the basis of CO titration experiments (55).

At an initial relative coverage of NO of 0.15, dissociation of the bridged NO is complete by 250 K, as shown by spectrum (b) of Fig. 6. If one assumes that the (coverage-independent) value of the pre-exponential factor of the rate coefficient for dissociation is 10^{13} s^{-1}, then the activation energy for dissociation of this state is approximately 16 kcal/mole. The coverage dependence of the activation barrier for dissociation of molecular NO is

illustrated by the data of Figs. 4 and 5. It has been inferred
from UPS data that there is such a coverage-dependent barrier for
dissociation of molecular NO on the (111) surface of Ni also (56).
Thermal desorption experiments indicate that there is a minimum relative coverage below which no desorption of molecular NO
occurs. Thermal desorption spectra of molecular $^{15}N^{16}O$ which
follow exposure to various amounts of $^{15}N^{16}O$ at 150 K are shown in
Fig. 8. The heating rate used in the desorption experiments was
21 ± 1 K/s. The maximum amount of $^{15}N^{16}O$ which desorbed represented only 5% of the total ^{15}N which desorbed as mass 30. (This
relative intensity may be subject to slight error due to prefer-
ential adsorption of $^{15}N^{16}O$ at the walls of the vacuum chamber.)
The molecular $^{15}N^{16}O$ desorbs as a single first-order peak at 480 K,
from which it follows that the value for the activation energy for
desorption is 31 kcal/mole, assuming that the pre-exponential
factor of the desorption rate coefficient is 10^{13} s^{-1}. Below
$\hat{\theta}_{NO}$ = 0.5, no molecular $^{15}N^{16}O$ desorbs. The onset of desorption
of molecular NO at intermediate relative coverages of NO has been
observed also on Ni(111) (56), polycrystalline Rh (55), Ru(100)
(8,10) and Ru(101) (11). In our experiments, desorption as molecular NO may occur only when the coverage-dependent activation
energy for dissociation becomes comparable to the activation energy for desorption. Another perspective is that desorption of some
molecular NO above $\hat{\theta}_{NO}$ = 0.5 may be necessary to free sites on the
surface required for dissociation of the remaining NO.

It is interesting to note the parallels which apparently
exist between the chemisorptive properties of NO on the Ru(001)
surface and on a dispersed supported Ru catalyst (57,58). Temperature programmed desorption from the reduced catalyst indicates
that, following exposure at 300 K, the low-coverage desorption
products result primarily from dissociation of NO; at higher coverages, desorption of molecular NO predominates (57). Furthermore,
complete oxidation of the catalyst leads to the exclusive adsorption and desorption of molecular NO (57). Based on these data, it
was concluded that initial adsorption of NO at 300 K is dissociative, and adsorption as molecular NO occurs only when the sites
required for dissociative adsorption have been blocked (57). The
infrared spectra of NO on supported Ru indicate also that oxidation
of the surface occurs initially, with absorption bands attributed to
adsorbed molecular NO occurring at 1810 – 1880 cm^{-1} (58), which may
be comparable to the high-frequency state at 1783 to 1823 cm^{-1}
observed on Ru(001). While the analogy between these two surfaces
is not complete, the similarities are certainly striking.

Co-adsorption of Nitric Oxide and Carbon Monoxide (13)

Co-adsorption of nitric oxide and carbon monoxide yields further insight into the adsorption of each of these molecules separately on the clean Ru(001) surface and the nature of their interaction when they are co-adsorbed on this surface. Previous LEED

data for CO on Ru(001) (38) indicate that adsorbed CO forms coin-
cidence lattices at absolute coverages, θ_{CO}, greater than 1/3, in
which many of the CO molecules are in low-symmetry adsorption
sites. A $(\sqrt{3} \times \sqrt{3})R30°$ LEED pattern is observed at $\theta_{CO} = 1/3$, however, and presumably corresponds to CO molecules adsorbed in sites
of high symmetry (37,38). The energy loss spectra for this system,
represented in Fig. 9, show a single C-O stretching frequency which
shifts with increasing coverage(13,59). Under the conditions of these
experiments, the frequency of the C-O vibration following a 10 L
exposure is 2033 cm^{-1}. An energy loss feature at approximately
444 cm^{-1} is assigned to the vibration of the CO molecule against
the Ru surface (Ru-CO stretching vibration), and the feature at
approximately 855 cm^{-1} is the first harmonic of this vibration. The C-O
vibration occurs in the frequency range commonly assigned to CO bound
linearly to a single metal atom, in spite of the observation of coinci-
dence lattices in the LEED structure. This is evidence that the Ru(001)
surface is electronically and geometrically homogeneous with respect
to adsorption of CO at high coverages, where CO-CO repulsive interactions predominate in the formation of the ordered overlayer (38,39).

In Fig. 10, vibrational spectra are shown for four experiments in which the Ru surface, with four different initial coverages of CO, was exposed to various amounts of NO. It is apparent
that for initial absolute coverages of CO, θ_{CO}^o, less than 0.3 [Fig.
10(A) and (B)], the subsequent adsorption of NO occurs with an
attenuation of intensity of the linear NO loss feature [1791 - 1831
cm^{-1} in Fig. 10(A) and (B)] relative to the loss spectra of NO on
clean Ru(001) which were obtained after equal NO exposures (Fig. 2).
Qualitative comparison of the intensities of the loss features in
Fig. 10(A) and (B) suggests that adsorption of NO may be accompanied by some displacement of CO from the surface for $\theta_{CO}^o \leq 0.3$. The
observed variations in energy of the vibrational loss features may
be due to changes in coverage of the individual adsorbates and/or
interactions between different adsorbates. At $\theta_{CO}^o \geq 0.3$, adsorption of NO occurs only in the bridged sites [1395 - 1460 cm^{-1} in
Fig. 10(C) and (D)]. The fact that NO can adsorb into the linear
sites only below $\theta_{CO}^o \sim 0.3$ is evidence that the CO molecules in the
$(\sqrt{3} \times \sqrt{3})R30°$ lattice are adsorbed in the on-top sites, as is the
case also for adsorption of CO on Pt(111) (60,61,62).

In Fig. 11, vibrational spectra are shown for three experiments in which exposure to NO was followed by exposure to CO, the
reverse of the exposure sequence of Fig. 10. The spectrum of Fig.
11(A) is to be compared with the spectrum of Fig. 2 which follows
a 120 torr-s NO exposure. It can be seen in Fig. 11(A) that exposure to CO causes displacement and/or conversion of the NO which
would have yielded the loss feature at 1783 cm^{-1} (linear NO) in the
absence of exposure to CO. In Fig. 11(B), the conversion/displace-
ment is obviously more difficult due to the higher initial coverage
of NO relative to the saturation coverage of NO on the clean surface $(\hat{\theta}_{NO} \equiv 1$ at saturation). Qualitative comparison of the relative peak intensities of Figs. 11(A) and (B) implies also that

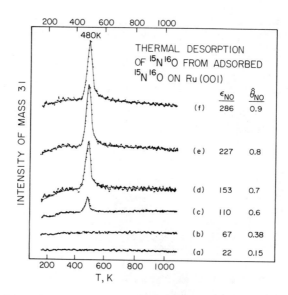

Figure 8. Thermal desorption spectra of $^{15}N^{16}O$ following adsorption of $^{15}N^{16}O$ on Ru(001) at 150 K. The baselines of the desorption spectra are displaced relative to one another for clarity.

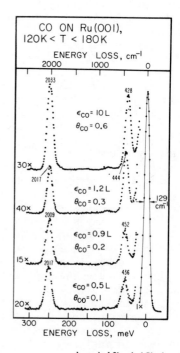

Figure 9. Electron energy loss spectra for various coverages (exposures) of CO on Ru(001) (13)

Journal of Chemical Physics

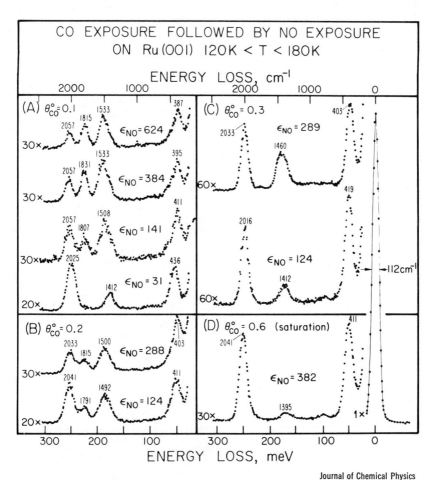

Journal of Chemical Physics

Figure 10.   Electron energy loss spectra of the Ru(001) surface following exposure to CO and subsequent exposure to NO at 120 K ≤ T ≤ 180 K (13)

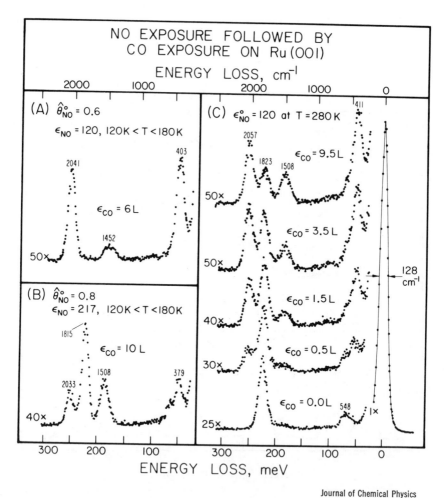

Journal of Chemical Physics

Figure 11.   Electron energy loss spectra of the Ru(001) surface following exposure to NO and subsequent exposure to CO at the temperatures indicated (13)

less CO adsorption occurs for the higher value of $\hat{\theta}^o_{NO}$.

Finally, the energy loss spectra of Fig. 11(C) show clearly that conversion from linear to bridged NO occurs during adsorption of CO. Whether displacement of NO by CO from the surface occurs also cannot be determined on the basis of these data. The surface represented by the spectrum at the bottom of Fig. 11(C) was prepared by exposing the Ru(001) crystal to NO at 280 K. At an exposure temperature between 120 K and 180 K, this 120 torr-s exposure would have resulted in the spectrum of Fig. 2, and a relative surface coverage of $\hat{\theta}_{NO} = 0.6$. Exposure at the higher temperature allows dissociation of bridged NO to occur, resulting in the exclusive population of the linear molecular NO state at this coverage. This observation is entirely consistent with the dissociation experiments discussed earlier. Exposure to CO of the surface of Fig. 11(C), where only linear NO and dissociation products are present, results in the growth of the 1508 cm^{-1} loss feature (bridged molecular NO), the growth of the 2057 cm^{-1} feature (adsorbed CO), and a decrease in the intensity of the 1823 cm^{-1} feature (linear molecular NO) relative to the other two. The 548 cm^{-1} feature (Ru-O and Ru-NO stretching vibrations) becomes almost obscured by the 411 cm^{-1} feature (Ru-CO vibration) as the CO exposure increases.

The observed competitive adsorption of CO and NO on Ru(001) has a close analogy in the reactions of two metal cluster compounds. The compound $Os_3(CO)_{12}$, which has 12 linear, singly-coordinated CO ligands, reacts with gaseous NO to form $Os_3(CO)_9(NO)_2$, in which three of the carbonyl ligands on one Os atom have been replaced by two singly-coordinated nitrosyl ligands. This compound can be reacted further with CO(g) to form $Os_3(CO)_{10}(NO)_2$, in which both nitrosyl ligands are now two-fold coordinated (bridged), and an Os-Os bond has been broken (63). The reaction sequence is illustrated in Fig. 12. This reaction proceeds similarly for the $Ru_3(CO)_{12}$ cluster (64). The conversion of the nitrosyl ligand by CO(g) from a single to a multiple coordination state with the metal is thus seen in both heterogeneous and homogeneous Ru systems.

It is interesting also to compare the results of the present experiment, which shows directly that a competitive mechanism occurs in the co-adsorption of NO and CO, with previous studies on several surfaces of the platinum group metals. On Pt(111) and Pt(110), Lambert and Comrie (65) have inferred from thermal desorption data that gaseous CO displaces molecular NO from the surface and causes also a conversion between two thermal desorption states of molecular NO. Similarly, Campbell and White (55) report that adsorbed CO inhibits the oxidation of CO by NO at low temperature on polycrystalline Rh. They attribute this to the occupation of sites by CO which are required for NO adsorption and dissociation. Conrad et al. (66) have used UV-photoelectron spectroscopy to observe directly the displacement of molecular NO by gaseous CO from Pd(110) and polycrystalline Pd surfaces. Thus, it appears that adsorption of molecular CO and NO is competitive on these

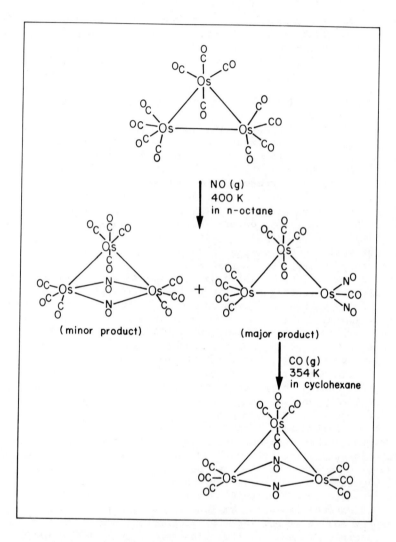

Figure 12.   Illustration of the reaction of $Os_3(CO)_{12}$ with NO(g) and subsequent reaction with CO(g) (13, 63)

surfaces also, at a temperature below that at which oxidation of CO occurs.

It has been suggested that an adsorbed cyanato complex (NCO) exists on supported Ru catalysts under conditions of relatively high pressure, on the basis of transmission infrared experiments (67,68). However, no evidence for formation of such an intermediate, which is expected to have vibrational frequencies at $1150 - 1480$ cm^{-1} due to the C-O stretch, and at $2150 - 2280$ cm^{-1} due to the N-C stretch (46), was observed in the co-adsorption experiments on the Ru(001) surface.

Co-adsorption of Nitric Oxide and Hydrogen (69)

When hydrogen alone is adsorbed on Ru(001), no new energy loss features are observable in the vibrational spectrum (69). A recent low-energy electron diffraction structural analysis has shown that, on Ni(111), hydrogen occupies three-fold hollow sites when it is in an ordered (2x2) superstructure at a half monolayer coverage (70). It has been deduced also from an analysis of "dipole-allowed" and "dipole forbidden" vibrations observed in EEL spectra that hydrogen occupies the hollow sites on the Pt(111) surface, with vibrational frequencies of 550 and 1230 cm^{-1}, at high coverage (33). By analogy, hydrogen might be expected to occupy the three-fold sites on the geometrically similar Ru(001) surface also. In this case, the Ru-H loss feature(s) may be unobserved due to a low dynamic dipole moment or to a low vibrational frequency.

In Fig. 13 are shown the vibrational spectra of a surface which was exposed to nitric oxide [Fig. 13(a)] to populate the bridged state, then to hydrogen [Fig. 13(b)]. The results of the experiment are identical when deuterium is used rather than hydrogen [Fig. 13(c)]. Thermal desorption peak areas indicate that 40-60% as much hydrogen is adsorbed on the surface of Fig. 13(b) as would be present in the absence of pre-adsorbed nitric oxide. The fact that the intensity of the loss feature which is caused by linear NO has grown, indicates that the hydrogen has displaced some of the bridged NO into the linear sites, thus supporting the hypothesis that hydrogen adsorption occurs in the three-fold hollow sites on Ru(001).

Conclusions

The results of the experiments discussed in this chapter may be summarized as follows:

(a) Nitric oxide adsorbs molecularly on Ru(001) at 150 K into two types of sites: a linear, "on-top" site, and either a twofold bridged or (more probably) a three-fold hollow site. The assignments of the EELS vibrational frequencies are consistent with infrared spectra of metal nitrosyl complexes. The relative population of these two states is a function of the total

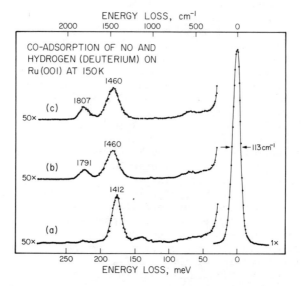

Figure 13.   Electron energy loss spectra of the Ru(001) surface following exposure to NO and subsequent exposure to H₂ (D₂) (69)

NO coverage, with the bridged state being populated exclus- ively at coverages below approximately a third of saturation.

(b)   The activation barrier for dissociation of adsorbed NO is coverage-dependent. At low relative coverages, the bridged NO dissociates with an activation energy of approximately 16 kcal/mole. This coverage-dependence is a consequence of site competition between molecular NO and its dissociation products.

(c)   At a sufficiently high coverage of N and O adatoms, the Ru(001) surface is poisoned with respect to subsequent ad- sorption of bridged NO.

(d)   Desorption of molecular NO begins to occur only at intermedi- ate coverages. This may be related to the coverage dependence of the activation energy for dissociation of bridged NO.

(e)   Carbon monoxide competes directly with NO for the adsites which are occupied by the linear form of NO. Exposure to CO causes at least partial conversion from the linear to bridged form of NO as adsorption of CO occurs.

(f)   The fact that some NO can adsorb into the linear sites only when the initial absolute coverage of CO is less than one- third is evidence that the CO molecules in the $(\sqrt{3} \times \sqrt{3})R30°$ superstructure are adsorbed in the on-top sites.

(g)   Hydrogen competes with NO for the adsites which are occupied by bridged NO.

Acknowledgments

It is a pleasure to acknowledge the contributions of our co- workers, Dr. Glenn E. Thomas and Dr. John T. Yates, Jr. This work was supported by the National Science Foundation under Grant No. CHE77-16314.

Literature Cited

1.   Dixon, J.K.; Longfeld, J.E. "Mechanism of Catalytic Oxidation" in "Catalysis" Vol. 7; Emmett, P.H., Ed.; Reinhold Publishing Corp.: New York, N.Y., 1960; p. 347.

2.   Thomas, C.L. "Catalytic Processes and Proven Catalysts"; Academic Press: New York, N.Y., 1970; p. 185.

3.   Klimisch, R.L.; Larson, J.G., Eds. "The Catalytic Chemistry of Nitrogen Oxides"; Plenum Press: New York, N.Y., 1975.

4.   Shelef, M.; Gandhi, H.S. Ind. Eng. Chem. Prod. Res. Dev., 1972, 11, 393.

5.   Klimisch, R.L.; Taylor, K.C. Envir. Sci. Technol., 1973, 7, 127.

6.   Kobylinski, T.P.; Taylor, B.W. J. Catal., 1974, 33, 376.

7.   Bonzel, H.P.; Fischer, T.E. Surface Sci., 1975, 51, 213.

8.   Ku, R.; Gjostein, N.A.; Bonzel, H.P. Surface Sci., 1977, 64, 465; see also p. 19 of Ref. (3).

9.   Orent, T.W.; Hanson, R.S. Surface Sci., 1977, 67, 325.

10.   Klein, R.; Shih, A. Surface Sci., 1977, 69, 403.

11. Reed, P.D.; Comrie, C.M.; Lambert, R.M. Surface Sci., 1978, 72, 423.

12. Thomas, G.E.; Weinberg, W.H. Phys. Rev. Letters, 1978, 41, 1181.

13. Thiel, P.A.; Weinberg, W.H.; Yates, J.T., Jr. J. Chem. Phys., 1979, 71, 1643.

14. Thiel, P.A.; Weinberg, W.H.; Yates, J.T., Jr. Chem. Phys. Letters, 1979, 67, 403.

15. Umbach, E.; Kulkarni, S.; Feulner, P.; Menzel, D. Surface Sci., 1979, 88, 65.

16. Cotton, F.A.; Wilkinson, G. "Advanced Inorganic Chemistry"; Interscience Publishers: New York, N.Y., 1972; p. 713.

17. Eisenberg, R.; Meyer, C.D. Acc. Chem. Res., 1975, 8, 26.

18. Enemark, J.H.; Feltham, R.D. Coord. Chem. Rev., 1974, 13, 339.

19. Kanski, J.; Rhodin, T.N. Surface Sci., 1977, 65, 63.

20. Zhdan, P.A.; Boreskov, G.K.; Boronin, A.I.; Schepelin, A.P.; Egelhoff, W.F., Jr.; Weinberg, W.H. J. Catal., 1979, 60, 93.

21. Ibach, H.; Lehwald, S. Surface Sci., 1978, 76, 1.

22. Pirug, G.; Bonzel, H.P.; Hopster, H.; Ibach, H. J. Chem. Phys., 1979, 71, 593.

23. Ibach, H. Surface Sci., 1977, 66, 56.

24. Newns, D.M. Phys. Letters, 1977, 60A, 461.

25. Šokčević, D.; Lenac, Z.; Brako, R.; Šunjic, M. Z. Phys. B., 1977, 28, 273.

26. Persson, B.N.J. Solid State Commun., 1977, 24, 573.

27. Evans, E.; Mills, D.L. Phys. Rev. B., 1972, 5, 4126.

28. Evans, E.; Mills, D.L. Phys. Rev. B., 1973, 7, 853.

29. Thomas, G.E.; Weinberg, W.H. J. Chem. Phys., 1979, 70, 1000.

30. Delanaye, F.; Lucas, A.; Mahan, G.D. Surface Sci., 1978, 70, 629.

31. Ho, W.; Willis, R.F.; Plummer, E.W. Phys. Rev. Letters, 1978, 40, 1463.

32. Willis, R.F.; Ho, W.; Plummer, E.W. Surface Sci., 1979, 80, 593.

33. Baró, A.M.; Ibach, H.; Bruchmann, H.D. Surface Sci., 1979, 88, 384.

34. Andersson, S.; Davenport, J.W. Solid State Commun., 1978, 28, 677.

35. Davenport, J.W.; Ho, W.; Schrieffer, J.R. Phys. Rev. B, 1978, 17, 3115.

36. Wendelken, J.F.; Withrow, S.P.; Foster, C.A. Rev. Sci. Instrum., 1977, 48, 1215.

37. Madey, T.E.; Menzel, D. Jpn. J. Appl. Phys., Suppl. 2, Part 2, 1974, 50, 229.

38. Williams, E.D.; Weinberg, W.H. Surface Sci., 1979, 82, 93.

39. Thomas, G.E.; Weinberg, W.H. J. Chem. Phys., 1979, 70, 954.

40. Thomas, G.E.; Weinberg, W.H. J. Chem. Phys., 1978, 69, 3611.

41. Sandstrom, D.R.; Withrow, S.P. J. Vac. Sci. Technol., 1977, 14, 748.

42. Goodman, D.W.; Madey, T.E.; Ono, M.; Yates, J.T., Jr. J. Catal., 1977, 50, 279.

43. "Dual Range Ionization Gauge Instruction Manual", No. 87-400257; Varian Associates: Palo Alto, CA, 1974; p. V-3.
44. Kuyatt, C.E.; Simpson, J.A. Rev. Sci. Instrum., 1967, 38, 103.
45. Thomas, G.E.; Weinberg, W.H. Rev. Sci. Instrum., 1979, 50, 497.
46. Nakamoto, K. "Infrared and Raman Spectra of Inorganic and Coordination Compounds"; John Wiley and Sons: New York, N.Y., 1978; p. 295.
47. Mercer, E.E.; McAllister, W.A.; Durig, J.R. Inorganic Chem., 1966, 5, 1881.
48. Poliakoff, M.; Turner, J.J. Chem. Commun., 1970, 1008.
49. Poliakoff, M.; Turner, J.J. J. Chem. Soc. A, 1971, 654.
50. Norton, J.R.; Collman, J.P., Dolcetti, G.; Robinson, W.T. Inorganic Chem., 1972, 11, 382.
51. Müller, J.; Schmitt, S. J. Organomet. Chem., 1975, 97, C54.
52. Toyuki, H. Spectrochim. Acta, 1971, 27A, 985.
53. Hoskins, B.F.; Whillans, F.D. J. Chem. Soc. Dalton, 1973, 607.
54. Quinby, M.S.; Feltham, R.D. Inorganic Chem., 1972, 11, 2468.
55. Campbell, C.T.; White, J.M. Appl. Surface Sci., 1978, 1, 347.
56. Conrad, H.; Ertl, G.; Küppers, J.; Latta, E.E. Surface Sci., 1975, 50, 296.
57. Uchida, M.; Bell, A.T. J. Catal., 1979, 60, 204.
58. Davydov, A.A.; Bell, A.T. J. Catal., 1977, 49, 332.
59. Thomas, G.E.; Weinberg, W.H. J. Chem. Phys., 1979, 70, 1437.
60. Froitzheim, H.; Hopster, H.; Ibach, H.; Lehwald, S. Appl. Phys., 1977, 13, 147.
61. Krebs, H.-J.; Lüth, H. Appl. Phys., 1977, 14, 337.
62. Hopster, H.; Ibach, H. Surface Sci., 1978, 77, 109.
63. Bhaduri, S.; Johnson, B.F.G.; Lewis, J.; Watson, D.J.; Zuccaro, C. J. Chem. Soc. Chem. Commun., 1977, 477.
64. Johnson, B.F.G.; private communication.
65. Lambert, R.M.; Comrie, C.M. Surface Sci., 1974, 46, 61.
66. Conrad, H.; Ertl, G.; Küppers, J.; Latta, E.E. Faraday Discuss. Chem. Soc., 1974, 58, 116.
67. Davydov, A.A.; Bell, A.T. J. Catal., 1977, 49, 345.
68. Brown, M.F.; Gonzalez, R.D. J. Catal., 1976, 44, 477.
69. Thiel, P.A.; Weinberg, W.H. J. Chem. Phys.,(in press) 1979.
70. Christmann, K.; Behm, R.J.; Ertl, G.; Van Hove, M.A.; Weinberg, W.H. J. Chem. Phys., 1979, 70, 4168.

RECEIVED July 31, 1979.

Inelastic Electron Tunneling Spectroscopy

JOHN KIRTLEY

IBM, Thomas J. Watson Research Center, Yorktown Heights, NY 10598

Inelastic electron tunneling spectroscopy (IETS) is a sensitive technique for obtaining the vibrational spectra of monolayer or submonolayer coverages of molecular layers at the interface region of metal-insulator-metal tunneling junctions. By intentionally doping the interface region with molecules of interest, the tunneling junction can be used as a model system for the study of chemisorption and catalysis on oxide and supported metal catalysts.   Jaklevic and Lambe (1,2) first described the phenomena of inelastic electron tunneling in 1966. Since that time nearly 100 papers, including 4 review articles (3,4,5,6) and a symposium proceedings (7), have been published on the subject. I will not attempt to comprehensively review all of this work here. Instead, I will try to present in a short, self-contained paper the basics of the subject: what IETS is, why it works, and a few of the things that can be done with it.   I must apologize in advance for slighting a great deal of excellent work in the field, both in the theoretical section, and especially in the applications section. The interested reader should refer to the reviews cited.

Section II will discuss the basic phenomena of inelastic tunneling from the viewpoint of the experimentalist. Section III will treat peak shapes, shifts, and widths.   Section IV will deal with intensities and selection rules in IETS. Finally, Section V includes some recent applications of IETS to the fields of chemisorption and catalysis, and to the at first glance unrelated field of surface enhanced Raman spectroscopy.

The Fundamentals

Inelastic electron tunneling junctions are prepared (Fig. 1) as follows: 1) A thin (~1000 Å) metal film is evaporated onto a clean insulating substrate in a vacuum system at ~10^{-6} torr. This film (the base electrode) is most often Al, although Pb (8),Sn (8) ,Cr (9), Ta (9), and Mg (2,10) have also been used successfully. This film is defined as a stripe ~1mm wide   by evaporation through an Al mask. 2) The film is oxidized by glow discharge anodization or by exposure to air or oxygen. Much effort has gone into finding reproducible methods for producing a good oxide, which must be thin (15-20 Å), pinhole free, and able to withstand high fields (typically ~10^6 volts/cm). 3) The oxide is doped with the materials of interest. Finely divided metals can be produced by evaporation (11). Molecular

0-8412-0585-X/80/47-137-217$07.25/0
© 1980 American Chemical Society

species are added by: exposing the films to a vapor of volatile materials (12), evaporating solids directly onto the base electrode (13), or dissolving the dopant, dropping the solution onto the films, and spinning off the excess in a photoresist type spinner (14). 4) The junctions are then completed by evaporating a counter-electrode through a second metal mask. Dopants can also be infused through the counter-electrode after completion of the junction (15,16), as will be described in Sec V.

Tunneling junctions are most often produced in a crossed stripe geometry so that 4-terminal measurements of their current-voltage characteristics can be made. Electrical contacts are made to the films (often with miniature brass "C" clamps), the samples are mounted in a Dewar insert, and cooled to liquid helium temperatures (4.2 ° K or below).

At these temperatures the distribution of occupied levels in the conduction bands (the Fermi distributions) in the two metal electrodes (Fig.1) are quite sharp, with a boundary between filled and empty states (the Fermi level) of characteristic width k_bT (k_b=0.08617 meV/K=0.69503 cm^{-1}/K). An applied bias voltage V between the two electrodes separates the Fermi levels by an energy eV. If the barrier oxide is sufficiently thin electrons can tunnel from one electrode to the other. This process is called tunneling since the electrons go through a potential barrier, rather than being excited over it. The barrier must be thin for an appreciable barrier to flow. For a typical 2 eV barrier the junction resistance is proportional to $e^{1.45s}$, where s is the barrier width in Angstroms (17). The tunneling current is predominantly elastic (no energy lost during the tunneling transition), but it is also possible for the electron to tunnel inelastically, losing energy hν to a barrier excitation. The initial electronic state must be filled and the final state must start empty for a tunneling transition to occur. At low temperatures this can only happen when eV$\geq$hν. Therefore an inelastic tunneling channel associated with an energy loss is switched on when the junction bias is increased.

The current-voltage (I-V) characteristic of a junction with only one possible excitation (Fig. 1) will show an increase in slope at eV=hν. This increase in slope is extremely small, hidden under a smoothly varying background, and is one of many possible excitations for a real tunneling junction. It is therefore necessary to take two derivatives of the I-V characteristic to bring out the inelastic structure and to reduce the elastic background. The barrier excitations then appear as peaks (Fig. 1) very similar to the absorption peaks observed in infrared absorption or the light scattering peaks observed in Raman scattering.

The derivatives are taken with a standard current modulation technique (Fig. 2). A slowly varying voltage sweep is applied across a large value resistor R_{dc} in series with the tunneling junction. If the resistance of the junction R_j is much less than R_{dc}, the current through the circuit is insensitive to the value of R_j, which varies as the applied bias increases. A small modulation current at ~1000 hz is added to the dc bias current through a large series resistor R_{ac}. The dc component of the resultant voltage across the junction is preamplified and applied to the X-channel of a chart recorder or digital computer. The resultant ac voltage across the junction is phase sensitively detected and applied to the Y-channel of the recording devices. We can expand the voltage across the tunneling junction in a Taylor series:

$$V(I)=V(I_0)+(dV/dI)(I-I_0) + (d^2V/dI^2)(I-I_0)^2/2+ \cdots \cdots \qquad (1)$$

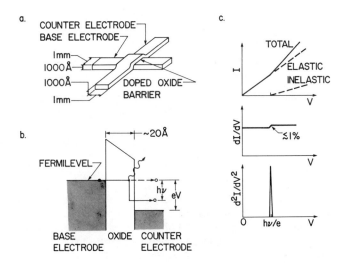

Figure 1. *(a) Tunneling junction geometry. A thin (∼ 1000Å) base electrode ∼ 1 mm wide is evaporated on an insulating substrate, oxidized to 15–20Å, doped with organic impurities, and a counter-electrode is evaporated to complete the junction. (b) Energy level diagram of the tunneling barrier region. An applied bias voltage V separates the Fermi levels by an energy eV, enabling electrons in filled states on one side of the barrier to tunnel into previously empty states on the other side. At low temperatures inelastic transitions can only occur for eV > hv, where hv is the energy loss associated with a particular vibrational excitation. (c) Schematic of the current–voltage characteristics of an inelastic tunneling junction with only one possible vibrational loss. The opening of the inelastic channel causes a slight increase in the conductance of the junction; the second derivative of the I–V characteristic shows a peak at eV = hv.*

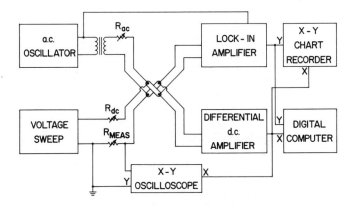

Figure 2. *Simplified schematic of the electronics involved in differentiating the I–V characteristic of a tunnel junction. The second harmonic voltage, proportional to d^2V/dI^2, is plotted vs. applied bias in a standard tunneling spectrum.*

For a current source ($R_{ac} \gg dV/dI$) we set $I\text{-}I_0 = I_\omega \cos(\omega t)$. By matching Fourier components we find that the first harmonic signal V_ω is proportional to the dynamic resistance of the junction:

$$V_\omega = (dV/dI)I_\omega \qquad (2)$$

and that the second harmonic signal $V_{2\omega}$ is proportional to the second derivative of the I-V characteristic times the square of the modulation current:

$$V_{2\omega} = -(d^2V/dI^2)I_\omega^2/4 \qquad (3)$$

Fig.3 shows the second harmonic signal vs applied bias for an undoped Al-AlO$_x$-Pb junction at 4.2 K with a modulation voltage of approximately 2mV. The dashed line is for the Pb counter-electrode superconducting, the solid line is for Pb normal (by applying a magnetic field of ~1000 Oe). This Figure illustrates the background signal in IETS. The major inelastic features are the aluminum oxide phonons (and their second harmonics), the vibrational modes of the hydroxyl ions in the bulk oxide and at the oxide surface, and the carbon-hydrogen stretching modes from residual impurities. In addition there is a smoothly varying background due to the change in the elastic conductance as the bias voltage is changed, and very sharp structure below ~200 cm^{-1} due to modification of the tunneling density of states by the Pb phonons. This last structure is hundreds of times larger than the inelastic tunneling structure when the Pb is superconducting. Quenching the superconductivity gives residual structure comparable in size to the inelastic structure. Vibrational structure as low in energy as ~100 cm^{-1} can be obtained by subtracting the second harmonic signal of an undoped junction from a doped junction using either an analog ac bridge technique (18) or by digital methods.

The tunneling spectrum of a doped junction can be seen in Fig. 4. In this case we have an Al-AlO$_x$-4-pyridine-carboxylic acid-Ag sample, with approximately monolayer coverage, run at 1.4 K. Fig. 4a shows the modulation (first harmonic) voltage V_ω across the junction as a function of applied bias. Since the modulation current I_ω is kept constant, V_ω is proportional to the dynamic resistance of the sample. The second harmonic voltage $V_{2\omega}$ (Fig. 4b), proportional to d^2V/dI^2 , shows the vibrational spectrum of the absorbed molecules. As we shall see below, a quantity which is more closely related to the density of vibrational oscillator strengths $D(\nu)$ is d^2I/dV^2. We show in Fig.4c the quantity

$$G_0^{-1}dG/d(eV) = (eG_0)^{-1}d^2I/dV^2 = -4V_\omega^{ref}V_{2\omega}/(V_\omega)^3 \qquad (4)$$

for this sample, where $G_0 = I_\omega/V_\omega^{ref}$ is the dynamic conductance of the junction at a reference voltage (in this case 200 mV). Integration of Eq. (4) over energy gives the ratio of the change in conductance to the total conductance due to a particular inelastic excitation. It is easier to identify the vibrational spectrum if the elastic background is subtracted out. We have done this in Fig. 4d by choosing a number of points where the inelastic structure is apparently absent, fitting these points to a fourth order polynomial expansion, and subtracting out the polynomial from the data.

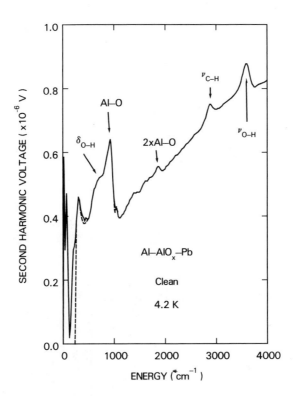

Figure 3.   Tunneling spectrum of an Al–AlO$_x$–Pb junction with no intentional dopants, for a 2 mV rms modulation voltage at 4.2 K: (– – –) superconducting electrode, (———) Pb normal. Inelastic structures that are present even in a junction with no intentional dopants, through aluminum phonons, bulk hydroxyl ions, surface hydroxyl ions, and residual hydrocarbon impurities, are pointed out. The elastic structure below ~ 200 cm^{-1} is hundreds of times larger than the inelastic structure for Pb superconducting, but comparable in size to the inelastic structure if the Pb superconductivity is quenched.

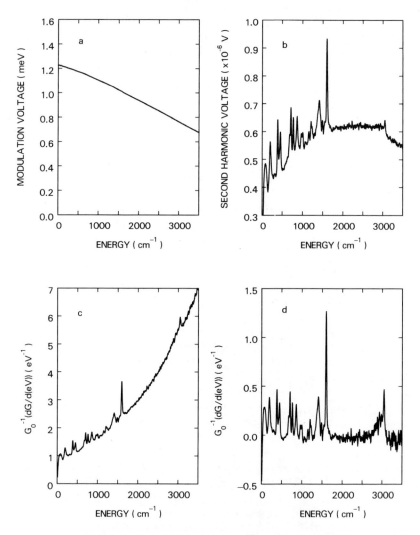

Figure 4.   Tunneling characteristics of an Al–AlO$_x$–4-pyridine–COOH–Ag junction run at 1.4 K with a ~ 1 mV modulation voltage. (a) Modulation voltage V$_\omega$ across the junction for a constant modulation current I$_\omega$. This signal is proportional to the dynamic resistance of the sample. (b) Second harmonic signal, proportional to d^{2}V/dI^2. (c) Numerically obtained normalized second derivative signal G$_o^{-1}$dG/ d(eV), which is more closely related to the molecular vibrational density of states. (d) Normalized G$_o^{-1}$dG/d(eV) with the smooth elastic background subtracted out numerically.

The vibrational spectrum of 4-pyridine-carboxylic acid on alumina in Fig. 4d is equivalent to an infrared or Raman spectrum and can provide a great deal of information about the structure and bonding characteristics of the molecular layer on the oxide surface. For example, the absence of the characteristic $\nu_{C=O}$ mode at 1680 cm^{-1} and the presence of the symmetric and anti-symmetric O-C-O stretching frequencies at 1380 and 1550 cm^{-1} indicate that 4-pyridine-carboxylic acid loses a proton and bonds to the aluminum oxide as a carboxylate ion.

Resolution and Sensitivity

If we neglect the energy dependence of the inelastic tunneling matrix elements within a particular band s, we can write for the inelastic tunneling current (2) :

$$I_i^s = K^s \int_0^\infty d\nu \ D^s(\nu) \int_{-\infty}^\infty dE \ f(E)(1 - f(e + eV - h\nu))N_1(E)N_2(E + eV - h\nu) \quad (5)$$

where K^s is a constant proportional to the matrix element for the sth inelastic transition, $D^s(\nu)$ is the vibrational density of oscillator strengths for the sth band at frequency ν, f(E) is the Fermi distribution of filled electronic states in the metal electrodes:

$$f(E) = (1 + e^{\beta E})^{-1} \quad (6)$$

where $\beta = (k_b T)^{-1}$, E is measured relative to the Fermi surface, and N(E) is the tunneling density of states, defined as N(E)=1 for a normal (non-superconducting) electrode, and

$$N(E) = |E|/(E^2 - \Delta^2)^{1/2} \quad |E| > \Delta, \quad (7)$$

N(E)=0 otherwise, for a superconducting electrode.

We will consider a junction with at most one superconducting electrode. Then we can set $N_2(E+eV-h\nu)=1$. Differentiating Eq. 5 with respect to voltage:

$$\frac{d^2 I_i^s}{dV^2} = K^s \int_0^\infty d\nu \ D^s(\nu)F(eV - h\nu) \quad (8)$$

$$F(eV - h\nu) = \int_{-\infty}^\infty dE \ f(E)(d^2 f(E + eV - h\nu)/dV^2)N_1(E) \quad (9)$$

where

$$d^2 f(x)/dV^2 = (\beta e)^2 e^x (1 - e^x)/(1 + e^x)^3 \quad (10)$$

If both films are normal ($N_1(E)=1$) the integration in equation (9) can be done analyically to obtain (2) :

$$F_n(\eta) = \beta e^2 e^\eta \left[\frac{(\eta - 2)e^\eta + (\eta + 2)}{(e^\eta - 1)^3} \right] \tag{11}$$

$$\eta \equiv \beta(eV - h\nu)$$

$F_n(\eta)$ is a bell shaped curve with full width at half maximum of 5.4 k_bT (Fig. 5a). For film 1 superconducting the integration of Eq. 9 must be done numerically. The results are shown in Fig. 5a for the ratio Δ/k_bT=3.0 (corresponding to Pb at ~4.2K) and for Δ/k_bT=12 (corresponding to Pb at ~1.2 K). The integrated area of all 3 curves of Fig. 5a are the same if the negative undershoots associated with the superconducting electrode are subtracted from the peaks. One can see from the curves that several things happen when one electrode goes from the normal to the superconducting state: 1) The peaks get higher and narrower. The full widths at half maximum for Δ= 0, 3k_bT, and 12k_bT are respectively 5.4k_bT, 2.9k_bT, and 2.8k_bT. 2) The peaks are shifted out in energy by roughly the gap energy. The peak shifts for Δ= 3k_bT and 12k_bT are respectively 0.80Δ and 0.94Δ. (The observed shifts are reduced somewhat by the finite modulation voltage used. A full treatment of the corrections to measured peak energies in the presence of superconductivity appears in (19)). 3). The lineshapes develop an undershoot.

Observed lineshapes in IETS are further modified by the finite modulation voltages used. We can write the second harmonic current signal at a bias voltage V_0 for a given non-linear I-V characteristic as (10) :

$$I_{2\omega}(eV_0) = \frac{2}{\tau} \int_0^\tau dt\, I(eV_0 + eV_\omega \cos(\omega t)) \cos(2\omega t) \tag{12}$$

Defining E=e$V_\omega \cos(\omega$ t) and integrating Eq. 5 by parts twice we obtain:

$$I_{2\omega}(eV_0) = K^s \int_{-\infty}^\infty de\, G(E) \frac{d^2 I(eV_0 + E)}{dV^2} \tag{13}$$

where

$$G(E) = \frac{2eV_\omega}{3\pi} (1 - (E/eV_\omega)^2)^{3/2} \quad |E| < eV,$$

$G(E)=0$ otherwise.

$G(E)$ (plotted in Fig.5b) is a bell shaped surve with full width at half maximum of 1.22 eV_ω .

Combining Eq.s (8) and (13) gives:

$$I_{2\omega}(eV_0) = K^s \int_{-\infty}^\infty de\, G(E) \int_0^\infty d\nu\, D^s(\nu)\, F(eV_0 + E - h\nu) \tag{14}$$

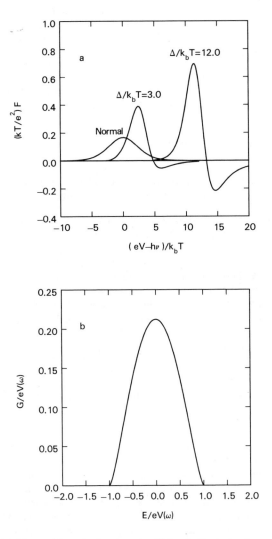

Figure 5. (a) Thermal broadening function F for both electrodes normal, and for one electrode superconducting with $\Delta/k_bT = 3.0$ and 12.0. When one electrode goes superconducting, the peaks shift out by $\sim \Delta$, become taller and narrower. The integrated area for all three curves are the same, taking into account the undershoots. (b) Modulation broadening function G. The observed IETS peaks are a double convolution of the vibrational density of states $D(\nu)$ with the thermal broadening function F and the modulation broadening function G.

The observed second harmonic signal is the double convolution of the oscillator density of states function $D^S(\nu)$ with a thermal broadening function $F(eV_0+E-h\nu)$ (Fig.5a) and a modulation broadening function $G(E)$ (Fig.5b).

The combined instrumental widths are plotted in Fig. 6 as a function of modulation voltage for two temperatures and for one electrode superconducting or both electrodes normal. The superconducting gap energy Δ was set equal to 1.2 meV, characteristic of Pb, for the superconducting case. The modulation voltages in Fig. 6 are r.m.s. since that is how they are measured experimentally. (The voltage modulation amplitudes V_ω are $\sqrt{2}$ larger than the r.m.s. voltages V_ω^{rms}. To convert to cm^{-1} multiply these numbers by 8.065). The total instrumental width for normal electrodes follows closely the expression:

$$\delta V_T = ((1.73 e V_\omega^{rms})^2 + (5.4 k_b T)^2)^{1/2} \qquad (15)$$

Fig. 6 shows that for a superconducting electrode the instrumental linewidths are dominated by modulation broadening for modulation voltages greater than 0.5 mV.

The ultimate resolution attainable using IETS appears to depend, up to a point, on the patience of the experimentalist. However, after that point the experimentalist must be very patient indeed. We can see from Eq. 13 that the observed signal in IETS is proportional to the square of the modulation voltage if the modulation voltage is less than the half-width of the structure in d^2I/dV^2. The instrumental widths are proportional to the modulation voltage at sufficiently low temperatures. Therefore, as in many spectroscopies, there is a tradeoff between signal/noise ratio and resolution. The signal due to a strong scatterer like 4-pyridine-carboxylic acid in monolayer coverage for a 1 mV r.m.s. modulation voltage (Fig. 4) is $\sim.35\mu$ V for the 1597 cm^{-1} line. The intrinsic voltage noise from a tunneling junction ([20]) is dominated by shot noise at the voltages of interest and is given by:

$$v_n/B_n^{1/2} = [2eI(V)R_j^2]^{1/2} \qquad (16)$$

where B_n is the bandwidth of the detector (inversely porportional to the output time constant.) If we assume an ohmic junction with $R_j = 100\Omega$, this works out to an intrinsic noise of 2.5×10^{-9} volts-$hz^{-1/2}$ at an applied bias of 200 mV (1613cm^{-1}). The measured noise level in our system is somewhat higher than this, $\sim 2 \times 10^{-8}$ volts-$hz^{-1/2}$, possibly because the lock-in pre-amplifier used (in our case a PAR 119 in direct mode) has a fairly high noise figure at the junction impedances and modulation frequencies used. At this noise level we could reduce the modulation to .6 mV (instrumental widths of ~ 8 cm^{-1}) and still retain a signal to noise ratio of 10 with a lock-in time constant of 3 sec. If the system were optimized to retain only shot noise, a modulation level of .2mV (instrumental width of 3 cm^{-1}) could be used. Of course, resolution could be improved still more by further reducing the modulation voltage and increasing the signal averaging time, but a full 0-3000 cm^{-1} scan with a 3 second time constant requires ~ 2 hours; scan times can quickly become prohibitively long.

High sensitivity, on the other hand, can be achieved in IETS by using larger modulation voltages. At 2 mV r.m.s. modulation the signal to noise ratio for a

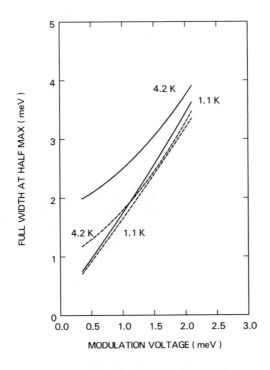

IETS INSTRUMENTAL LINEWIDTHS

Figure 6. Full width at half-maximum of the IETS spectral lines, for an infinitely narrow vibrational density of states, vs. modulation voltage, for two temperatures, and for the counter-electrode (——) normal or (– – –) superconducting

monolayer of 4-pyridine-carboxylic acid with a 3 second time constant is ~100. Therefore one could expect to detect .05 monolayers with a signal/noise ratio of 5 using the same modulation voltage and time constant. There is some danger in this kind of extrapolation, since intensities do not vary exactly linearly with surface coverage in IETS (21) (as we will see below), but Kroeker and Hansma (22) have reported distinguishing one molecular species from a second in a 1%-99% mixture.

If we are interested in using tunneling junctions as model systems for chemisorption and catalytic reactions, it is important to know how strong an effect the counter-electrode has on the system. Real catalysts are not coated with thick layers of Pb. Kirtley and Hansma measured the vibrational frequencies of hydroxyl ions (23) and benzoate ions (19) on alumina using several different counter-electrodes, and compared these measurements with infrared measurements of the same species on high surface area alumina samples, with of course no counter-electrode. They found that: 1) vibrational frequencies were always lower when measured using IETS than when measured using optical methods, 2) different counter electrodes caused different vibrational mode shifts, and 3) these shifts could be modeled empirically with a simple image dipole model. Their results are summarized in Fig. 7, which plots the measured vibrational frequencies (in meV) against $1/R^3$, where R is the atomic radius of the counter-electrode metal in angstroms. The simple image model predicted vibrational mode shifts proportional to $(Ze)^2$, where Ze is the dipole moment derivative associated with a particular mode, and predicted shifts inversely proportional to d^3, where d is an effective charge-image plane distance. The shifts are relatively large (~4%) for the stretching vibrations of the hydroxyl ions (which have a large Z~0.3); small (<1%) for the C-H bending modes of the benzoate ions, (which have Z~0.1); and immeasurably small for the ring deformations, which might be expected to have large d's. The important point is that the vibrational mode shifts due to the counter-electrode are systematic and not sufficiently large either to interfere with identification of the modes by comparison with optical measurements, or to indicate that the surface chemistry is significantly different in the presence of a counter-electrode.

Intensities and Selection Rules

The first theoretical treatment of intensities in IETS was presented by Scalapino and Marcus (24). They included the modification of the tunneling potential barrier by a simple molecular dipole potential to calculate the ratio of the inelastic tunneling current to the elastic tunneling current, and predicted approximately correct absolute magnitudes. Jaklevic and Lambe (2) extended this treatment to include the interaction of the tunneling electrons with the polarizability of the molecules. Kirtley, Scalapino, and Hansma (25) (KSH) modified the approach of Scalapino and Marcus to include multiple image effects and the effects of nonspecular transmission. Several other approaches to the calculation of intensities in IETS have been presented. In particular some of the formal difficulties with KSH, including the use of Bardeen's transfer Hamiltonian formalism (26) and the WKB approximation for the electronic wavefunctions (27,28), have been avoided. We will limit our discussion to KSH because of space limitations, and because KSH is the only theory to date that has been used to calculate intensities for complex molecules.

As mentioned above, KSH used Bardeen's transfer Hamiltonian formalism (29): an interaction potential transferred electrons from initial states on one side of

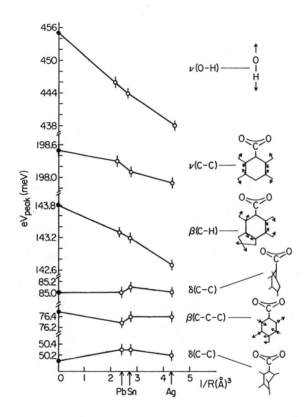

Physical Review B

Figure 7.   Measured vibrational energies for hydroxyl and benzoate ions on alumina using (●) optical techniques and (○) IETS for three different counterelectrode metals. The energies measured using IETS are always lower than those measured optically, but the shifts are not large enough to interfere with identification of the modes or to indicate large perturbations of the surface species by the presence of the counterelectrode (19).

the tunneling junction to final states at a lower energy on the other side. The interaction potential had the form of a sum of dipole potentials localized on the individual atoms in the molecule. All of the images in the two metal surfaces, as well as dielectric screening in the oxide, were included in the interaction potential. The initial and final electronic wave functions were written for a square barrier potential using the WKB approximation (30). KSH found that when appropriate values for the barrier parameters (obtained by fitting the total junction I-V characterisitics) and molecular dipole derivatives were used, approximately correct absolute intensities were predicted. Further, they found that Raman active as well as infrared active modes should be observed using IETS, even neglecting the effects of molecular polarizabilities. Observation of both infrared and Raman active modes have indeed been observed in a surface layer of anthracene on alumina using IETS. The relaxation of dipole symmetry selection rules occured because of spatial inhomogeneities in the inelastic tunneling matrix elements. The spatial inhomogeneities normal to the interface were due to boundary conditions imposed by the metal surfaces; spatial inhomogeneities parallel to the interface were due to constructive and destructive interference of the matrix elements caused by oscillations of the electronic wavefunctions

In order to have confidence in a simple dipole potential calculation, it was important to know the relative range of the interaction; whether long or short range forces were important. KSH showed that the long range nature of the inelastic tunneling interaction could be demonstrated by measuring the ratio of inelastic peak heights for opposite junction bias potentials. Since the molecules were doped onto the surface of the oxide, the electrons lost energy before tunneling through most of the barrier for one bias potential, and lost energy after tunneling through most of the barrier for the opposite bias. More energetic electrons were more likely to penetrate the barrier, and therefore one bias direction was favored over the other. A short range potential caused large asymmetries in peak amplitudes; a long range potential caused smaller aymmetries. In fact, KSH showed that the dipole interaction fit the opposite bias asymmetries of peak heights for benzoate ions on alumina fairly well, but that a point interaction predicted asymmetries much too large. These conclusions have been supported recently by measurements for a number of molecules by Muldoon, Dragoset, and Coleman (31).

Further support for a long range interaction in IETS came from calculations by Kirtley and Soven (32) showing that multiple resonance scattering of the initial and final electronic wavefunctions, which would be expected for a short range interaction, lead to large second harmonic peaks, much larger than are observed experimentally.

Near one metal electrode the potential for a dipole oriented normal to the interface tends to add with that of its image in the metal electrode. The potentials of the dipole and its image tend to cancel for a dipole oriented parallel to the interface. The addition or cancellation of the dipole potential with its image makes it more probable for an inelastic transition to occur for a dipole oriented normal to the interface. Near the center of a tunneling junction, however, the reverse is true for two reasons. 1) The cancellation of potentials for the dipole oriented parallel is least important at the center. 2) The inelastic transition matrix element involves an integration over the barrier volume of the interaction potential times some nearly spatially homogeneous electronic wavefunction terms. If the dipole is oriented normal to the interfaces and located in the center of the barrier, the potential is an odd function in z (if z defines the normal to the interfaces), and integrates to zero. Fig. 8 plots the inelastic tunneling cross section against position for a point dipole in a 15Å thick barrier for 3 different vibrational energy losses, and for dipoles oriented normal (solid line) and parallel (dashed curve) to the interface. Fig. 8 shows that vibrational excitations normal to the interface are indeed favored if the molecule is close to the metal electrode, as is believed to be the case, but that some caution should be used when invoking an orientation "selection rule". If caution is indeed exercised, it may be possible to determine the orientation of molecules on surfaces by analyzing IETS intensities.

A test of this possibility came from an analysis of the IETS intensities of methyl sulfonic acid on alumina. Hall and Hansma (33) used the vibrational mode energies of this surface species to show that it was ionically bonded to the alumina and that the SO_3^- group (with tetrahedral bonding) had oxygen atoms in nearly equivalent chemical positions. They predicted that the molecule, which had a surface geometry of two back to back tripods, was oriented with the C-S bond normal to the oxide surface.

The IETS intensities for the methyl group vibrations of this species are shown in Fig. 9. The theoretical predictions of Kirtley and Hall (34) using KSH, and taking methyl group dipole derivatives from infrared measurements of ethane, assuming the C-S bond normal vs parallel to the interface, are also shown in Fig. 9. Note that for an orientation with the C-S bond normal, the symmetric C-H modes (2 and 9), which have net dipoles parallel to the C-S bond, are favored over the anti-symmetric modes (4,7, and 11), which have net dipole moments perpendicular to the C-S bond, but that for the C-S bond parallel to the surface the situation is reversed. The better, although by no means perfect, agreement between theory and experiment for the C-S bond normal tends to support the proposed orientation of Hall and Hansma.

Nearly all theories to date predict that IETS intensities should be proportional to n, the surface density of molecular scatterers. Langan and Hansma (21) used radioactively labeled chemicals to measure a surface concentration vs solution concentration curve (Fig. 10) for benzoic acid on alumina using the liquid doping technique. The dashed line in Fig. 10 is a 2 parameter fit to the data using a simple statistical mechanical model by Cederberg and Kirtley (35). This model matched the free energy of the molecule on the surface with that in solution. The two parameters in this model were the surface density of binding sites ($\sim 10^{-1}$Å^{-2}) and a parameter related to the binding energy per site (~ 1 eV). The solid line in Fig. 10 shows the intensity of the 686 cm^{-1} line of the benzoate ion as a function of solution concentration. The striking feature of this data is that the two curves are not parallel; the tunneling intensity is not proportional to the surface concentration, but goes approximately as n$^{1.4}$.

Cunningham, Weinberg, and Hardy (36) proposed that cooperative behavior between the adsorbed molecules results in an intensity vs surface coverage curve that is non-linear, with a power law exponent falling between 1.1 and 1.65. A second possible explanation is that tunneling occurs in the thin spots between dopant molecules at low coverages, decreasing the relative inelastic current because of the larger effective distance between the tunneling electron and the surface species, and increasing the elastic current. It is not clear at this point which view is more nearly correct. More work on this problem needs to be done.

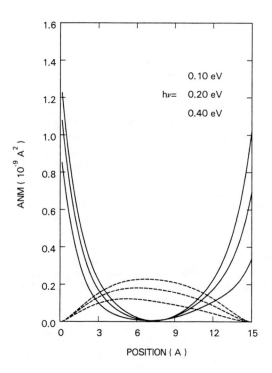

MATRIX ELEMENT SQUARED

Figure 8.   Inelastic tunneling cross section, from the theory of Kirtley, Scalapino, and Hansma (KSH), for a point dipole as a function of position in a 15Å thick barrier, for 3 different vibrational energy losses. One electrode is taken to be at 0Å, the other at 15Å: (——) dipole oscillating normal to the interface, (– – –) dipole parallel to the interface. Dipoles oriented normal to the interfaces are favored close to the metal electrodes; dipoles oriented parallel to the interface are favored near the center of the barrier.

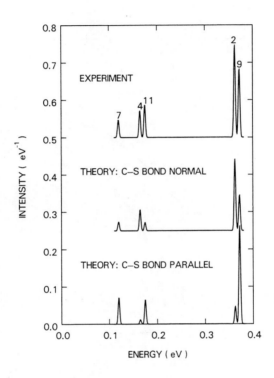

CH$_3$SO$_3$ IETS METHYL SPECTRAL BANDS

Figure 9.   Experimental and theoretical (KSH) IETS intensities for the methyl group vibrations of methyl sulfonate ions on alumina. The theoretical curves assume dipole derivatives from IR measurements of C_2H_6, and compares the predictions for the C–S bond normal vs. parallel to the interface. The relatively better fit for C–S normal supports the proposed orientation of Hall and Hansma.

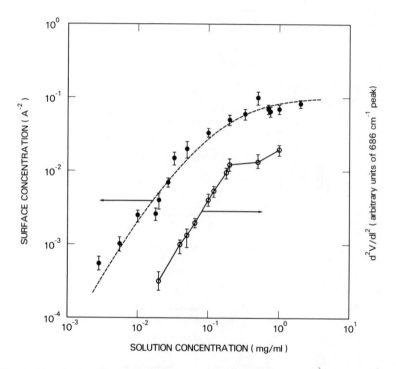

*Figure 10.   Surface concentration (●) and peak intensities (○) vs. solution con-
centration using the liquid doping technique for benzoate ions on aluminum: (– – –)
2-parameter fit to the surface concentration data proposed by Cederberg and Kirtley (25). The nonlinear dependence of IETS intensities on surface concentration
can be seen by noting that the two curves are not parallel. Experimental data from
Langen and Hansma (12).*

Applications

We will concentrate on four recent applications of IETS, chosen for the variety of problems and techniques they represent. This section is not intended to be exhaustive.

McBride and Hall (37,38) reported the first observation of a controlled catalytic reaction on alumina using IETS. They studied the catalytically induced transfer hydrogenation from water vapor to unsaturated hydrocarbon chains chemisorbed on alumina at both ends of the chain. They absorbed muconic acid (trans-trans-1,3   butadiene   1,4   dicarboxylic   acid,
HOOC-CH=CH-CH=CH-COOH) onto oxidized aluminum strips using the liquid doping technique. The samples were returned to the vacuum system, and in the presence of 0.3 torr of D_2O vapor, heated to up to 400° C by passing current through a heater strip evaporated on the back of the glass slide. The films were then allowed to cool and the junctions completed by evaporation of the Pb counter electrode.

The tunneling spectrum of unheated samples (Fig. 11a) showed that both ends of the muconic acid chemisorbed ionically to the oxide. Heated samples showed a monotonic increase in the C-D stretching mode intensity with temperature (Fig. 11b). Detailed comparisons of the spectra of muconic acid and its saturated counterpart, adipic acid ($HOOC-(CH)_4-COOH$), showed that the growth of the C-D stretching mode intensity was due to the hydrogenation of the chain rather than exchange with the hydrogens already in the molecule. The intensity of the ν_{C-D} peak depended exponentially on $1/T$, where T was the heater temperature (Fig. 12), was independent of the heating time from 3 to 45 minutes, and depended only weakly on the pressure of D_2O in the system.

These results indicated that the hydrogenation reaction used hydrogen (or deuterium) from the surface hydroxyl groups, that the number of hydroxyl groups available to the reaction depended on the temperature of the substrate, and that the reaction stopped when the available hydrogen was used up. The results of McBride and Hall emphasized the importance of surface hydroxyl groups in this catalytic reaction on alumina.

Jaklevic (16) showed that catalytic reactions could be observed proceeding after completion of the junction as well as before. He infused propiolic acid into a clean junction using a technique developed by Jaklevic and Gaerttner (39): the completed undoped junction was exposed to a near saturation vapor of a weak solution (2 to 10 % by weight) of the acid in water. The tunneling spectrum of the sample immediately after infusion (Fig. 13a) corresponded to the salt of propiolic acid. When such a junction was stored at room temperature for several weeks or heated in air to about 175° C for several minutes, the spectrum of Fig 13b appeared. Comparison with the infrared spectrum of sodium acrylate and with the tunneling spectrum of acrylic acid showed that the hydrogenation of the pro-pionate ion to the acrylate ion (illustrated at the bottom of Fig.13) was occuring. This reaction took place sufficiently slowly that tunneling spectra could be taken at intermediate times. All spectra could be interpreted as being due to linear combina-tions of the propionate and acrylate spectra, indicating that no intermediate state was being observed.

Many catalysts of commercial importance are highly dispersed metals support-ed on silicon or aluminum oxides. Hansma, Kaska, and Laine (11) showed that these catalysts could be modeled with tunneling junctions by evaporating very thin

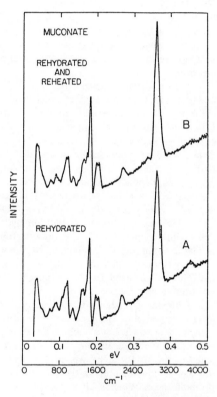

Figure 11. Tunneling spectrum of the muconate ion ($^-OOC—CH$=$CH—CH—CH—COO^-$) chemisorbed to alumina and heated to 50°C and 325°C in the presence of D_2O vapor. The spectrum changes continuously to that of the adipate ion ($^-OOC—(CH_2)^4—COO^-$), the saturated counterpart of the muconate ion. Deuterium from the D_2O vapor contributes heavy protons to the process, causing a growth of the ν_{C-D} peak at 2150 cm^{-1} (37).

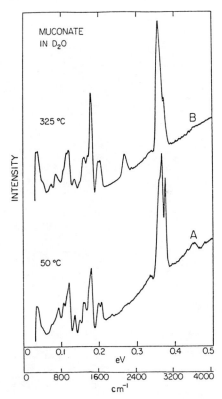

Figure 12. Plot of the intensity of the v_{C-D} peak shown in Figure 11 as a function of the temperature to which the Al film is heated. The reaction proceeds only to a certain point for a given temperature, indicating that the number of hydroxyl ions taking part in the hydrogenation reaction depends on temperature (37).

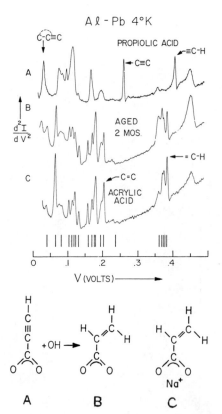

Figure 13. Tunneling spectra of propiolic acid on alumina immediately after infusion and after 2 months aging at room temperature, and compared with the tunneling spectrum of acrylic acid. The close agreement between the lower two tunneling spectra and the bar spectrum of sodium acrylate indicate that the hydrogenation reaction, shown schematically at the bottom of the figure, is ocurring in the junction after completion. Redrawn from (28).

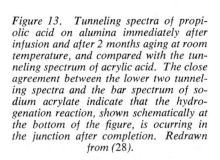

layers (~4 Å equivalent mass) of transition metals onto aluminum oxide. The metals formed highly dispersed 30 Å diameter balls on the surface of the oxide. Immediately after evaporation of Rh on alumina, Kroeker, Kaska, and Hansma (40,15) exposed the highly dispersed Rh metal particles to a saturation coverage of CO, and completed the junctions by evaporation of a Pb counter-electrode. The junctions were then transferred to a high pressure cell, and exposed to high pressures (~ 1×10^7 Pa) of H_2 at high temperatures (up to 440° K). The Pb electrodes were highly permeable to hydrogen, but not to residual impurities in the pressure cell, allowing the maintanence of a relatively clean surface under extreme conditions.

Analysis of the vibrational spectra of the junctions before exposure to hydrogen (Fig. 14) (including isotopic substitutions) showed the presence of at least three different CO species on the surface, two different linear-bonded species and at least one bridge bonded species.

Analysis of the tunneling spectra of the hydrocarbons formed by exposing the samples to H_2 (Fig. 14) showed two different species, one a formate like ion, the other an ethylidene ($CHCH_3$) species. The formate ion is not thought to be an active intermediate in hydrocarbon synthesis, but the ethylidene species may well be a catalytic intermediate. Kroeker, Kaska and Hansma were then able to suggest a reaction pathway for the hydrogenation of CO on a supported rhodium catalyst consistent with the formation of ethylidene as a catalytic intermediate.

Recent reports of large enhancements of the Raman scattering from monolayers of pyridine on Ag in electrochemical cells (41,42) (factors of 10^4-10^6 larger cross sections than for molecules in solution) have attracted considerable interest. One of the difficulties in understanding the mechanism (or as is commonly believed, mechanisms) producing these large cross sections is the complexity of the Ag -water interface in an electrochemical cell. For example, large enhancements only occur in an electrochemical cell after several atomic layers of Ag have been dissolved and replated onto the electrode (43). Controversy continues over whether this step is required to clean the Ag surface (43), roughen the surface on an atomic scale (44), roughen the surface on an optical scale (45), or all three.

Tsang and Kirtley (46) showed that the surface enhanced Raman effect also occured in tunneling junction structures. These structures had the advantage that some of the parameters present in electrochemical cells, such as roughness, could be varied independently of one another. They doped oxidized aluminum films with 4-pyridine-carboxylic acid and evaporated 200 Å thick films of Ag to complete the tunneling junction structure. The acid chemisorbed to the oxide as an ion, losing a proton, and leaving the pyridine group free to interact with the Ag. The Ag films were electrically continuous, so that tunneling spectra of the samples could be run, but also optically transparent, so that Raman scattering from the dopant-Ag interface was possible.

Tsang and Kirtley found that the Raman scattering cross section from a monolayer of 4-pyridine-carboxylic acid was enhanced by a factor of ~10^4 in the tunneling junction geometry over that in solution. Fig. 15 shows a comparison of the tunneling and Raman spectra of 4-pyridine-carboxylic acid with an Ag top electrode. Particularly suggestive are the relatively weak intensities of the Raman signal at low vibrational energies when compared with the IETS signal. The differences and similarities between the spectra obtained by these two methods on the same system may help our understanding of the mechanism for enhanced Raman scattering.

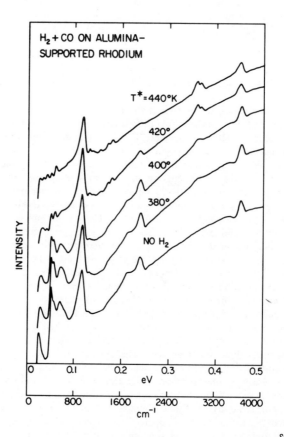

Springer–Verlag

Figure 14.   Tunneling spectra of a sample with finely dispersed Rh particles on alumina, exposed to a saturation coverage of CO, and heated to various temperatures in a high pressure atmosphere of H_2. The CO is hydrogenated on the surface. Analysis of the resultant spectra using isotopic substitution indicates that an intermediate species, ethylidene di-rhodium, is formed (7).

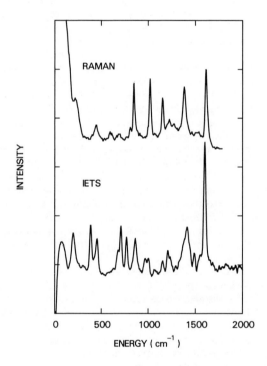

Figure 15. Comparison of the tunneling and Raman spectra of an Al–AlO$_x$–4-pyri-dine–COOH–Ag junction. The Raman cross section for this monolayer coverage is enhanced by a factor of 10^4 over solution cross sections. The different intensity patterns for the two spectroscopies may provide clues to the Raman enhancement mechanism.

The ability to characterize the interface region of these samples using IETS before taking a Raman spectrum proved very useful. For example, samples doped with benzoic acid showed strong IETS spectra, and therefore had molecules present in near monolayer coverage, but had undetectably small Raman scattering signals; identically prepared samples doped with 4-pyridine-COOH, with very similar IETS spectra, had large Raman signals. This indicated that the interaction between the particular dopant molecule and the Ag counter-electrode was important. Similarly, samples with monolayer coverages of 4-pyridine-carboxylic acid showed large Raman signals when the counter- electrode was Ag, but showed only weak signals for a Cu counter-electrode, and undetectably small signals for Pb,Au,and Al.

Tsang, Kirtley, and Bradley (45) also found that rough tunneling junctions gave larger Raman signals than smooth ones. The junctions were roughened by evaporating CaF_2 on the substrates before evaporating the Al bottom electrodes. The CaF_2 formed shallow mounds ~500 Å in diameter. The mounds grew higher as more CaF_2 was evaporated. Fig. 16 shows the Raman spectra of 3 samples with different thicknesses of CaF_2. The samples with.larger roughness amplitudes had larger Raman scattering signals. Since light couples strongly to surface plasmons only for rough surfaces, these results suggested that light coupled to surface plasmons in the Ag, causing the observed Raman enhancements.

A good demonstration that surface plasmons were indeed playing a role came from samples with Al-AlO$_x$-4-pyridine-COOH-Ag tunneling junctions on diffraction grating substrates (45). Light reflected off the gratings were absorbed by surface plasmons at scattering angles such that the grating periodicity matched the change in wavevector due to the absorption of a photon by a surface plasmon. In their multi-film geometry this occured at two angles, resulting in two dips in the reflectivity of the grating (Fig. 17) . The Raman scattering had peaks at the same angles (Fig. 17), showing convincingly the connection between surface plasmons and surface enhanced Raman scattering.

One simple explanation for these results was as follows: The electric field at a metal vacuum interface can be >10 times larger than in free space when the conditions required for a surface plasma resonance are met (47). Since the Raman cross-section is proportional to the square of the field, surface plasmons could produce enhancements of $>10^2$. This enhancement is probably not large enough to explain the tunneling junction results by itself, but an enhancement in signal of a factor of 100 by the excitation of surface plasmons would increase the Raman intensity from near the limits of detectibility.

VI. Conclusions

The applications of inelastic tunneling presented in Sec. V point up both the strong and weak points of this spectroscopy. Inelastic electron tunneling is sensitive, has good resolution, does not require large capital investment, has a wide spectral range, is sensitive to all surface vibrations, and can be used on oxide and supported metal catalysts. However, a counter-electrode must be used, single crystal metal surfaces cannot be used, and spectra must be run at low temperatures.

The necessity for a counter-electrode in IETS may be a handicap in some applications, but it can also be used to advantage. McBride and Hall let the surface reaction they were studying proceed to a certain point, quenched it, and then completed the tunneling junctions and ran their spectra. Jaklevic, and Kroeker, Kaska, and Hansma allowed the catalytic reaction to proceed within the junction,

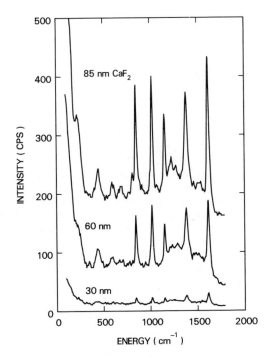

Physics Review Letters

Figure 16. Raman scattering signal from Al–AlO$_x$–4-pyridine–COOH–Ag tunneling junctions with three different thicknesses of CaF$_2$ evaporated on the substrates before making the samples (45). Thicker CaF$_2$ films have larger roughness amplitudes. The rougher Ag films couple better to surface plasmons and give larger Raman scattering enhancements.

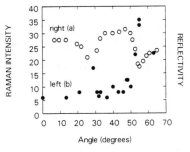

Physics Review Letters

Figure 17.   Reflectivity (○) and Raman scattering intensities (●) for an Al–AlO$_x$–4-pyridine–COOH–Ag tunneling junction prepared on a diffraction grating substrate, as a function of the angle between the incident laser beam and the normal to the grating surface (45). At the same angles that absorption by surface plasmons causes reflectivity dips, the Raman signal shows peaks.

taking advantage of the surprisingly high and selectable permeability of the top electrode. Tsang and Kirtley took advantage of the presence of the counter electrode to study the enhancement of Raman scattering from the dopant-Ag interface.

Inelastic electron tunneling spectroscopy has been shown to be a useful method for the study of chemisorption and catalysis on model oxide and supported metal catalyst systems. There are in addition a number of proven and potential applications in the fields of lubrication, adhesion (48), electron beam damage (49,50), and electrochemistry for the experimentalist who appreciates the advantages and limitations of the technique.

Abstract

Inelastic electron tunneling spectroscopy (IETS) is a sensitive technique for obtaining the vibrational spectra of monolayer or submonolayer coverages of molecular layers at the interface region of metal-insulator-metal tunneling junctions. By intentionally doping the interface regions with molecules of interest, the tunneling junctions can be used as model systems for the study of chemisorption and catalysis on oxide and supported metal catalysts. This paper presents the basics of the subject: what IETS is, why it works, and a few of the things that can be done with it. Included are: the basic phenomena; peak shapes, shifts,and widths; intensities and selection rules; and some recent applications to chemisorption and catalysis, and to the at first glance unrelated field of surface enhanced Raman spectroscopy.

Literature Cited

1. Jaklevic,R.C.;Lambe,J.;**Phys. Rev. Lett.**,1966,**17**,1139.
2. Lambe,J.;Jaklevic,R.C.;**Phys. Rev.**,1968,**165**,821.
3. Hansma,P.K.;**Phys. Rep.**,1977,**30C**,146.
4. Weinberg,W.H.;**Ann. Rev. Phys. Chem.**,1978,**29**,115.
5. Keil,R.G.;Roenker,K.P;Graham,T.P;**Appl. Spect.**,1976,**30**,1.
6. Hansma,P.K.;Kirtley,J.R.;**Acc. Chem. Res.**,1978,**11**,440.
7. Wolfram,T.,Ed." Inelastic Electron Tunneling Spectroscopy ";Springer-Verlag:Berlin;1978.
8. Yanson,I.K.;Bogatina,N.I.;**Sov. Phys. JETP**,1971,**32**,823.
9. McBride,D.;Rochlin,G.;Hansma,P.;**J. Appl. Phys.**,1974,**45**,2305.
10. Klein,J.;Leger,A;Belin,M,;Defourneau,D.;Sangster,M.J.L.;**Phys. Rev. B**,1973,**7**,2336.
11. Hanmsa,P.K.;Kaska,W.C.;Laine,R.M.;**J. Am. Chem. Soc.**,1976,**98**,6064.
12. Lewis,B.F.;Mosesman,M.;Weinberg,W.H.;**Surface Science**,1974,**41**,142.
13. Simonsen,M.G.;Coleman,R.V.;**Nature**,1973,**244**,218.
14. Simonsen,M.G.;Coleman,R.V.;Hansma,P.K.;**J. Chem. Phys.**,1974,**61**,3789.
15. Kroeker,R.M.;Kaska,W.C.;Hansma,P.K.;preprint to be published.
16. Jaklevic,R.C.;preprint to be published
17. A good introduction to tunneling can be found in: Burstein.E.;Lundqvist,S.; " Tunneling Phenomena in Solids ",1968,Plenum,New York.
18. Colley,S.;Hanmsa,P.K.;**Rev. Sci. Instr.**,1977,**48**,1192.
19. Kirtley,John;Hansma,Paul K.;**Phys. Rev. B**,1976,**13**,2910.
20. Moody,M.V.;Paterson,J.L.;Ciali,R.L.;**Rev. Sci. Instr.**,1979,**7**,903.
21. Langan,J.D.;Hansma,P.K.;**Surf. Sci.**,1975,**52**,211.
22. Wolfram,T.,Ed.,op.cit.,p.16.

23. Kirtley,J.R.;Hansma,P.K.;**Phys. Rev. B**,1975,**12**,531.
24. Scalapino,D.J.;Marcus,S.M.;**Phys. Rev. Lett.**,1967,**18**,459.
25. Kirtley,John;Scalapino,D.J.;Hansma,P.K.;**Phys. Rev. B**,1976,**14**,3177.
26. Feuchtwang,T.E.;**Phys. Rev. B**,1979,**20**,430.
27. Brailsford,A.D.;Davis,L.C.;**Phys. Rev. B**,1970,**2**,1708.
28. Davis,L.C.;**Phys. Rev. B**,1970,**2**,1714.
29. Bardeen,J.;**Phys. Rev. Lett.**,1961,**6**,57.
30. see, for example, Schiff,L.I.;" Quantum Mechanics ",McGraw-Hill:New York;1968
31. Muldoon,M.F.;Dragoset,R.A.;Coleman,R.V.;**Phys. Rev. B**,1979,**20**,416.
32. Kirtley,John;Soven,Paul;**Phys. Rev. B**,1979,**19**,1812.
33. Hall,J.T.;Hansma,P.K.;**Surf. Sci.**,1978,**71**.
34. Kirtley,J.R.;Hall,J.T.;submitted to **Phys. Rev. B**.
35. Cederberg,A.A.;Kirtley,J.R.;**Solid State Comm.**,1979,**30**,381.
36. Wolfram,T.,Ed.,op.cit.,p.125.
37. McBride,D.E.;Hall,J.T.;**J. Catalysis**,1979,**58**,320.
38. McBride,D.E.;Hall,J.T.;submitted
39. Jaklevic,R.C.;Gaerttner,M.R.;**Appl. Phys. Lett.**,1977,**30**,646.
40. Wolfram,T.,Ed.,op.cit.,p.186.
41. Fleischman,M.P.;Hendra,P.J.;Mcquillen,J.;**J. Chem. Soc. Chem. Commun.**,1973,**3**,80.
42. Jeanmaire,D.L.;van Duyne,R.P.;**J. Electroanal. Chem.**,1977,**82**,329.
43. Pettinger,B.;Wennig,U.;Kolb,D.H.;**Ber. Bunsenges. Phys. Chem.**,1978,**82**,329.
44. Otto,A.;**Proceedings. Int. Conf. Vibr. Absorbed Layers**,June 1978,Julich.
45. Tsang,J.C.;Kirtley,J.R.;Bradley,J.A.;**Phys. Rev. Lett.**,1979,**43**,772.
46. Tsang,J.C.;Kirtley,J.R.;**Solid State Commun.**,1979,**30**,617.
47. Philpott,M.R.;**J. Phys. Chem.**,1975,**61**,1812.
48. White,H.W.;Godwin,L.M.;Wolfram,T;**J. Adhesion**,1978,**9**,237.
49. Hanmsa,P.K.;Parikh,M.;**Science**,1975,**188**,1304.
50. Parikh,M.;Hall,J.T.;Hansma,P.K.;**Phys. Rev. A**,1976,**14**,1437.

RECEIVED June 3, 1980.

Neutron Scattering Spectroscopy of Adsorbed Molecules

H. TAUB

Department of Physics, University of Missouri–Columbia, Columbia, MO 65211

In this paper we discuss how neutron scattering spectroscopy can be applied to the study of the structure and dynamics of adsorbed molecules. Since reviews of elastic and inelastic neutron scattering from adsorbed films have recently appeared (1-3), our purpose here is not to present a comprehensive survey of every adsorbed system investigated by neutron scattering. Rather, we shall be concerned primarily with two questions which are basic to the characterization of adsorbed species on catalysts and which have been central to the discussion of this symposium. These are the extent to which the neutron scattering technique can be used to determine 1) the orientation and position of an adsorbed molecule; and 2) the strength and location of the forces bonding a molecule to a surface.

We begin by considering the distinguishing features of the neutron as a surface probe, comparing it to other surface spectroscopies. Although an attempt will be made to relate experiments discussed to previous work, we shall focus on those systems most favorable for study by neutron scattering and for which the greatest effort toward answering the above questions has been made. Thus, even though it is not of catalytic interest, the model system of n-butane adsorbed on graphite will be used to illustrate the capabilities and limitations of the technique. We then consider the progress which has been made in applying neutron scattering to molecules adsorbed on catalytic substrates. The most intense activity in this area has been with hydrogen chemisorbed on transition-metal substrates. The paper concludes with prospects for the future based on both the model system results and the experiments with transition-metal catalysts.

I. Distinguishing Features of Neutron Scattering as a Surface Probe

Probably the most distinguishing feature of the neutron as a surface probe is its penetrability. As an uncharged particle the

0-8412-0585-X/80/47-137-247$08.50/0
© 1980 American Chemical Society

neutron interacts weakly with matter and therefore does not scatter preferentially from surfaces. This lack of surface sensitivity limits neutron scattering to the study of films adsorbed on
high-surface-area substrates. Typically, substrates with surface
area ≥ 20 m^2/g are required. For elastic diffraction studies of the
structure of adsorbed monolayers these surface areas have been
sufficient to study a wide variety of adsorbates (1-3). However,
in the case of vibrational spectroscopy, the inelastic neutron
cross-sections are generally one to two orders of magnitude
smaller so that, with rare exceptions, one is limited to the study
of hydrogenous films.

There are, of course, a large number of high-surface-area
substrates. Charcoals, graphites, zeolites, metal oxides, Raney
nickel, and transition-metal powders have all been used in neutron
scattering experiments. Most of these have surface areas well
over 20 m^2/g and none are strong neutron absorbers. The problem,
then, is not one of obtaining adequate scattered intensities but
of having a sufficiently well-characterized surface to allow a
detailed interpretation of the experiment. By well-characterized
we mean a substrate consisting of crystalline particles having one
crystal plane as the primary adsorbing surface. In addition, for
elastic diffraction from adsorbed monolayers, it is necessary to
have particles of dimension ≥ 50 Å so that the coherence length of
the film will be sufficiently long to give well-defined Bragg
reflections. Probably the best characterized substrates currently
in use are graphitized carbon powders and recompressed exfoliated
graphites such as Grafoil and Papyex. (The trade names of some of
the more commonly used graphitized carbon powders are Graphon,
Carbopack B, Vulcan III, and Sterling F.T. Papyex is a recompressed exfoliated graphite made by Le Carbone Lorraine and is
very similar to Grafoil marketed by Union Carbide). Almost the
entire surface area of these substrates is provided by exposed
graphite basal planes of exceptional uniformity. In addition, the
exfoliated graphites possess preferred orientation of the basal
plane surfaces. The width of Bragg peaks observed by elastic
neutron scattering from monolayers adsorbed on Grafoil imply
coherence lengths of the film >100 Å. At least at the present
time, these graphite substrates offer better characterized surfaces albeit lower surface areas than commercially available
catalysts.

While limiting the technique to high-surface-area substrates,
the penetrability of the neutron has some compensating advantages
for studies of the adsorbed films. Unlike electron surface
spectroscopies, ultra-high vacuum is not required. Films can be
examined at equilibrium with their own vapor using samples similar
to those on which thermodynamic measurements such as vapor pressure isotherms and specific heats can be performed. In this way
elastic neutron diffraction has been useful in elucidating phase
diagrams of simple gases physisorbed on exfoliated graphite sub-

strates (1,3). Although not yet exploited, there is the possibility of performing neutron scattering experiments under high
ambient pressures such as those found in a catalytic reactor.

A second advantage related to the weakness of the neutron-
nuclear interaction is that multiple scattering effects can be
neglected in many experiments with adsorbed molecules. This
permits interpretation of both elastic and inelastic scattered
intensities in terms of well-known scattering laws (4). We shall
see that in the case of elastic neutron diffraction from monolayer
films it is possible to interpret the relative intensity of Bragg
reflections directly in terms of a geometrical structure factor.
In some cases this allows the orientation of the adsorbed molecules to be determined in addition to the two-dimensional unit
cell of the monolayer. This situation should be contrasted with
that for low-energy electron diffraction (LEED) where multiple
scattering effects greatly complicate the interpretation of Bragg
peak intensities from an overlayer.

As we shall discuss below, it is also more straightforward to
calculate the relative intensity of vibrational modes observed by
inelastic neutron scattering than in electron-energy-loss and
optical spectroscopies. The relative intensity of the modes, as
well as their frequency, can then be used to identify the atomic
displacement pattern or eigenvector of the mode. We shall also see
through examples of model calculations how the relative intensity
of surface vibratory modes is sensitive to the orientation of the
adsorbed molecule and the strength and location of its bond to the
surface.

Another feature of the neutron-nuclear interaction, in addition to its weakness, is that it does not exhibit any systematic
variation through the periodic table. While electron and photon
spectroscopies are most sensitive to atoms with large electron
concentrations, neutron spectroscopy is perhaps unique in its
effectiveness in probing the motion and position of hydrogen
atoms. Historically, this feature has allowed neutron scattering
to play an important role in both chemical crystallography and
molecular spectroscopy. The sensitivity to hydrogen is equally
important, if not more so, in studies of adsorbed molecules. The
spin-incoherent cross-section (4,5) of hydrogen is one to two
orders of magnitude greater than for most other atoms so that
scattered intensities are sufficient to study molecular vibrations
in submonolayer hydrogenous films. The scattering from hydrogen
will dominate that from other surface atoms present as well as the
much larger number of substrate atoms. Moreover, deuterium has a
far lower total neutron cross-section than hydrogen. Therefore,
deuteration of a molecule can yield not only a frequency shift of a
vibrational mode due to the mass increase but also a large decrease
in the mode intensity. On the other hand, for structural studies
of hydrogenous films by elastic neutron scattering, it is more
favorable to deuterate the molecule because of the larger coherent

cross-section of deuterium.

The dominance of the incoherent scattering from hydrogen together with the neglect of multiple scattering greatly simplifies the inelastic scattering law for hydrogenous species. The differential cross-section for one-phonon, incoherent, inelastic neutron scattering into solid angle Ω and with energy transfer $\hbar\omega$ is given by ([6](#))

$$\frac{\partial^2\sigma}{\partial\Omega\partial\omega} = \frac{k}{k_0} \frac{1}{N} \sum_{j,n} \left(\frac{\sigma_H^{inc}}{4\pi}\right) \exp(-2W_n)\exp(\hbar\omega/2k_BT)$$

$$\times \ \hbar(\vec{Q} \cdot \vec{C}_j^n)^2/4N\omega_j M_H \mathrm{csch}(\hbar\omega_j/2k_BT)\delta(\omega_j-\omega)$$

(1)

where σ_H^{inc} is the incoherent cross-section of hydrogen, M_H is the hydrogen mass, N is the number of hydrogen atoms in the molecule, ω_j and $\vec{C}_j^n$ are the frequency and hydrogen displacement vectors for the nth hydrogen atom in the jth normal mode, $\exp(-2W_n)$ is the Debye-Waller factor for the nth hydrogen, $\vec{k}_o$ and $\vec{k}$ are the incident and scattered momenta, and $\vec{Q} = \vec{k} - \vec{k}_o$ is the momentum transfer.

The cross-section in Eq. (1) illustrates another distinguishing feature of inelastic neutron scattering for vibrational spectroscopy, i.e., the absence of dipole and polarizability selection rules. In contrast, it is believed that in optical and inelastic electron surface spectroscopies that a vibrating molecule must possess a net component of a static or induced dipole moment perpendicular to a metal surface in order for the vibrational transition to be observed ([7](#),[8](#)). This is because dipole moment changes of the vibrating molecule parallel to the surface are canceled by an equal image moment induced in the metal.

While all vibrational transitions are allowed by Eq. (1), the intensity of a mode is governed by the $(\vec{Q}\cdot\vec{c}_j^n)^2$ term which expresses the component of the neutron momentum transfer along the direction of the atomic displacements. To an extent, this feature can be exploited with substrates such as Grafoil which have some preferred orientation. By aligning $\vec{Q}$ parallel or perpendicular to the predominant basal plane surfaces, the intensity of the "in-plane" and "out-of-plane" modes, respectively, can be enhanced. In practice, while this procedure can be useful in identifying modes ([9](#)), the comparison with calculated intensities can be complicated by uncertainties in the particle-orientation distribution function. In this respect, randomly oriented substrates are to be preferred ([10](#)).

II. Vibrational Spectroscopy of Adsorbed Molecules by Inelastic Neutron Scattering

A. Early experiments. The first inelastic neutron scattering experiments were performed about 15 years ago. This work has been reviewed by Boutin et al (11) and will not be discussed extensively here. The experiments investigated methane (12,13), ethylene (12,14), and hydrogen and acetylene (12) adsorbed on activated charcoal. The importance of these experiments lies in their recognition that sufficient scattered intensities could be obtained to study the dynamics of submonolayer hydrogenous films on high-surface-area substrates. Although evidence of diffusive motion was found, the experiments were unsuccessful in observing well-defined vibrational modes of the adsorbed molecules. This was probably due to the heterogeneity of the charcoal substrates employed and the relatively high temperatures at which the experiments were performed.

B. More recent experiments on physisorbed monolayers. A series of experiments with simple molecules physisorbed on graphite substrates have demonstrated that well-defined vibrational excitations can be observed in monolayer films by inelastic neutron scattering. The first of these studied argon monolayers adsorbed on Grafoil (9). As a monatomic, classical rare gas, argon was chosen for its simplicity. Furthermore, it was possible to use the isotope ^{36}Ar which has the largest coherent neutron cross-section of any element to enhance the ratio of film-to-substrate scattering. A schematic diagram of the triple-axis neutron spectrometer used in these experiments is shown in Fig. 1.

Elastic neutron diffraction was first performed (analyzer in Fig. 1 set to zero energy transfer) to establish the structure of the monolayer at low temperature. Three Bragg reflections were observed which could be indexed by a triangular lattice having a nearest-neighbor distance about 10% smaller than required for a $\sqrt{3}$ X $\sqrt{3}$ R30° commensurate structure (every third carbon hexagon in the graphite basal plane occupied).

Inelastic scattering measurements were then performed on this solid phase at 5 K in two different sample configurations: the direction of the momentum transfer $\vec{Q}$ parallel ("in-plane") and perpendicular ("out-of-plane") to the preferred orientation of the graphite basal planes. The inelastic spectra at two fixed values of Q = 2.75 Å^{-1} and 3.5 Å^{-1} are shown in Fig. 2(a) for the Q-parallel configuration. The two energy-loss peaks were identified as the excitation of tranverse ($\Delta E \simeq 3$ meV (24 cm^{-1})) and longitudinal ($\Delta E \simeq 5.5$ meV (44 cm^{-1})) zone boundary phonons of the 2D polycrystalline film. Calculations of the monolayer lattice dynamics assuming an Ar-Ar Lennard-Jones potential with nearest-neighbor interactions, but neglecting the interaction with the substrate, yielded good agreement with the observed phonon frequencies. The Q-dependence of the "in-plane" spectra results from the 2D polycrystalline average of the $(\vec{Q} \cdot \vec{c}_j)^2$ term which appears in the one-phonon, coherent, inelastic neutron cross-section (as in the incoherent cross-section of Eq. (1)).

The inelastic spectra obtained for the ^{36}Ar monolayer in the "out-of-plane" configuration are shown in Fig. 2(b) for the same

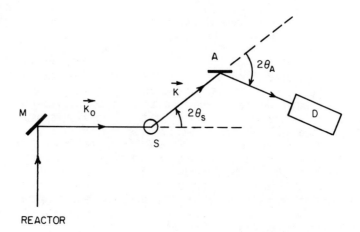

Figure 1.   Schematic of a triple-axis inelastic neutron spectrometer. The thermal-
ized beam from the reactor is monochromated by Bragg reflection from crystal M.
Neutrons that scatter from the sample S are energy analyzed by Bragg reflection
from crystal A and enter detector D.

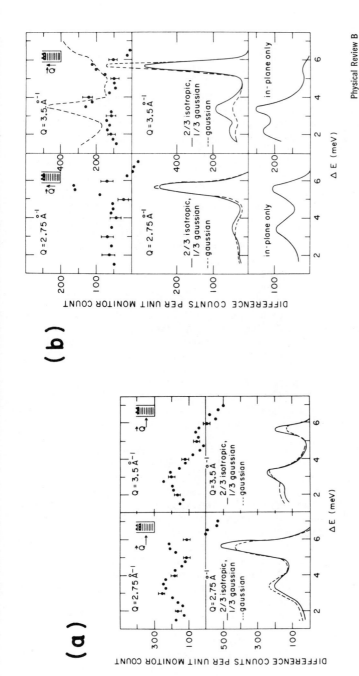

Physical Review B

Figure 2.   Inelastic neutron spectra from ^{36}Ar *monolayers adsorbed on Grafoil at 5 K (9). Curves plotted immediately below data are computed spectra for two different particle-orientation distribution functions: (a) "in-plane" configuration with momentum transfer Q parallel to the preferred orientation of the graphite basal planes; (b) "out-of-plane" configuration with Q perpendicular to the preferred basal planes.   Curves at the bottom represent the calculated contribution to the observed spectra caused by in-plane scattering from misoriented crystallites.*

fixed momentum transfers. The relative intensity of the peaks is qualitatively different than that for the "in-plane" configuration. The enhanced intensity at 5.6 meV (45 cm^{-1}) at both Q's was attributed to a dispersionless Einstein oscillator mode in which the argon atoms move normal to the surface. As noted above and shown at the bottom of Figs. 2(a) and 2(b), the relative intensity of the calculated modes depends on the particle orientation distribution function assumed for the substrate. It should be kept in mind that the distribution of particle orientations causes both in-plane and out-of-plane atomic motions to be excited in each sample configuration.

Although the question of orientation order does not arise, these results for argon monolayers illustrate several features of neutron vibrational spectroscopy. Well-defined excitations can be observed at somewhat lower energy transfers than accessible with optical and electron energy-loss spectroscopies. Typical energy resolution in the scans of Fig. 2 is ~0.3 meV (2.4 cm^{-1}). The capability of obtaining inelastic spectra at constant non-zero Q is also not present in these other spectroscopies. However, scattered intensities are relatively low. Counting times in Fig. 2 were ~30 minutes per point.

Very similar inelastic neutron scattering experiments using a triple-axis spectrometer have been performed on H_2 and D_2 monolayers adsorbed on Grafoil ([15]). Unlike argon, these films form a $\sqrt{3} \times \sqrt{3}$ R30° structure at low coverages. In the "out-of-plane" configuration, this commensurate phase had one-phonon energy-loss peaks at 4 meV (32 cm^{-1}) for D_2 and 5 meV (40 cm^{-1}) for H_2. The frequencies of these peaks were Q-independent so that, as with argon, the modes were attributed to the motion of an Einstein oscillator normal to the surface. At higher coverages the films became incommensurate and the inelastic spectrum exhibited broader features.

The rotational transitions of the H_2 molecule were also observed in monolayer films adsorbed on Grafoil ([15]). A sharp energy-loss peak appears at 14.6 meV (117 cm^{-1}) corresponding to the transition from the J = 0 (para) to the J = 1 (ortho) rotational state. The para-ortho conversion energy is about the same as for bulk hydrogen. From the narrow width of the excitation, it was concluded that the J = 1 line is not split into a doublet corresponding to J_z = 0 and a J_z = ±1 and hence that the ortho molecules are freely rotating. This experiment is one of the first applications of inelastic neutron scattering to infer orientational properties of an adsorbed molecule. A later experiment with hydrogen adsorbed on activated Al_2O_3 found a spectrum with a broader peak in the range 5-7 meV (40-55 cm^{-1}) which was interpreted to result from severely hindered orientational motion ([16]).

As an example of the extreme energy resolution possible with inelastic neutron sattering, we include some recent experiments of Newbery et al ([17]) on methane adsorbed on graphitized carbon blacks. Using a time-of-flight spectrometer and an incident neutron wavelength of 10 Å, they achieved an instrumental resolution of 20 μeV (0.16 cm^{-1}). Figure 3 shows the incoherent neutron scattering spectrum of 0.7 layers of CH_4 adsorbed on Vulcan III at several temperatures. Two side peaks are observed at 58 μeV (0.46 cm^{-1}) and 108 μeV (0.864 cm^{-1}) in both neutron energy loss and energy gain. These two transitions are ascribed to quantum mechanical tunneling of the molecule through a rotational barrier about axes parallel to the surface. The broad contour underlying both elastic and tunneling peaks is attributed to rotational diffusion about an axis <u>perpendicular</u> to the surface (<u>18</u>). The CH_4 molecule was assumed to be sitting like a tripod on the graphite basal planes with three hydrogen atoms close to the surface.

The symmetry and height of the rotational barriers and hence the tunnel splittings depend strongly on the orientation of the molecule and its adsorption site. The results of these measurements (a higher resolution experiment is planned) when combined with model calculations based on empirical atom-atom potentials (see below) may be able to provide corroborative evidence for the orientation of an adsorbed molecule as well as details of the molecule-substrate interaction. The principal obstacle to a wider application of the technique may simply be the small number of adsorbed systems in which these tunnel splittings can be observed.

C. <u>A model system for neutron vibrational spectroscopy of adsorbed molecules: monolayer butane on graphite</u>. In this section we concentrate on a particular system, monolayer <u>n</u>-butane adsorbed on graphite, for which a considerable effort has been made to analyze the inelastic neutron spectra for the orientation of the adsorbed molecule and the forces bonding it to the substrate (<u>10,19,20</u>). By treating one system in greater detail, we can better illustrate the capabilities and limitations of the technique.

Before discussing the adsorbed butane vibrational spectrum, we first describe the inelastic neutron spectrometer used in this experiment.

1. <u>Description of a time-of-flight inelastic neutron spectrometer.</u> A schematic diagram of the neutron time-of-flight (TOF) spectrometer at the University of Missouri Research Reactor Facility used in the adsorbed butane experiments is shown in Fig. 4. Most chemical applications of neutron vibrational spectroscopy use variations of this basic design. Neutrons originating in the reactor core pass through a refrigerated beryllium filter which rejects all fast neutrons with wavelength less than 4 Å. The remaining low-energy neutrons (E < 5 meV (40 cm^{-1})) are incident upon the sample. Those neutrons which either scatter elastically or gain energy by annihilating a phonon in the sample are detected. The energy of the scattered neutrons is analayzed by a TOF method in which the chopper after the sample creates a burst of neutrons whose flight time to the detector bank 5.8 m away can be measured.

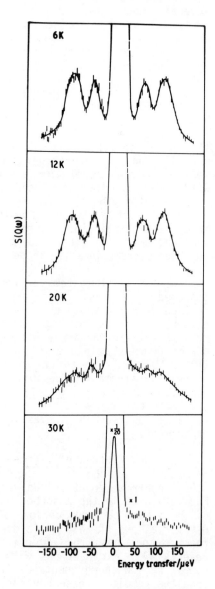

Figure 3. Tunneling spectra of 0.7 lay-ers CH_4 on a graphitized carbon powder Vulcan III at various temperatures (17). Incident neutron wavelength is 10Å and energy resolution is 20 μeV (0.16 cm^{-1}).

Chemical Physics Letters

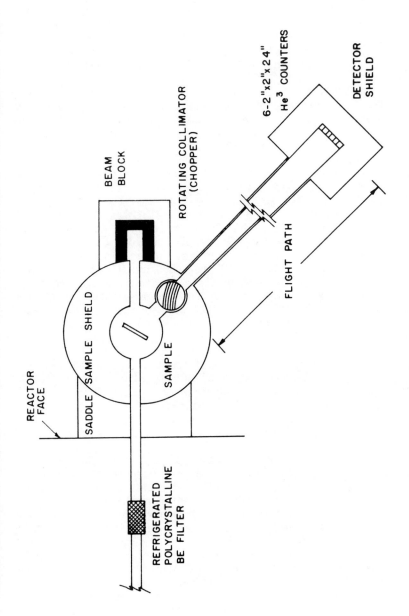

Figure 4. Schematic of the inelastic neutron time-of-flight spectrometer at the University of Missouri Research Reactor Facility

The detector is set at the smallest available scattering angle, 27.4°, to minimize Doppler broadening of the vibrational bands and multiphonon contributions to the inelastic scattering which vary as Q^4 (compare the Q^2 dependence in the one-phonon scattering of Eq. (1)). Since the observed intensity of the one-phonon modes depends upon their thermal population, at 77 K the useful range of energy transfers obtainable with this spectrometer is from about 6 meV (50 cm^{-1}) to 40 meV (320 cm^{-1}) with a resolution of ~10 cm^{-1}. The momentum transfer varies from 1 Å^{-1} to 3.5 Å^{-1} over this energy range. The fact that the inelastic spectra are not taken at fixed Q as is possible with a triple-axis spectrometer (Fig. 1) is of no concern if the excitations of interest have negligible dispersion.

Some of the alternative TOF instrument designs involve replacing the beryllium filter with either a crystal or a mechanical chopper to monochromate the incident beam. With this change, the spectrometer can be used with a higher incident neutron energy (typically $E_0 \sim$ 50 meV) so that a smaller momentum transfer Q is possible for the same energy transfer (21,22). With a monochromatic incident beam, a beryllium filter is sometimes substituted for the chopper after the sample in order to increase the scattered intensity but with a sacrifice in the minimum Q attainable. Energy transfers up to ~100 meV (800 cm^{-1}) can be achieved with TOF spectrometers at steady state reactors before the incident neutron flux is limited by the thermal spectrum of the reactor. (With hot moderators such as at the Institut Laue-Langevin reactor in Grenoble, useful neutron fluxes can be obtained up to ~0.5V (4000 cm^{-1})).

2. Butane monolayer vibrational spectrum and model calculations. The vibrational states of butane adsorbed on graphite have been studied in a series of experiments (10,19,20) at several coverages and temperatures. The substrate employed has been a graphitized carbon powder known as Carbopack B which was selected over Grafoil because of its larger surface area (~80 m^2/g vs ~30 m^2/g) yet high homogeneity.

Figure 5 contains a schematic diagram of the n-butane molecule ($CH_3(CH_2)_2CH_3$). While it may be regarded as too large a molecule to be taken as a model system, butane has several advantages for study by inelastic neutron scattering. Recall that it is the incoherent scattering from the hydrogen which contributes most strongly to the scattering. Therefore, in spite of the large number of internal degrees of freedom of the molecule, the bulk inelastic neutron spectrum is dominated by only three modes which involve large amplitude torsional motion of the methyl (CH_3) and methylene (CH_2) groups. As we shall see below, another advantage of butane is that in the monolayer vibrational spectrum the lowest lying internal torsion at ~150 cm^{-1} is well separated from a broader band centered at 50 cm^{-1}. This feature provides a convenient "window" for observing additional modes which result from the adsorbed molecule vibrating against the surface.

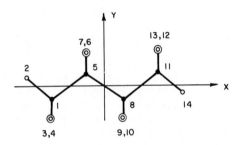

*Figure 5.   Schematic of the butane molecule: (○) hydrogen atoms; (●) carbons.
Atoms 1, 2, 5, 8, 11, and 14 are coplanar. The concentric circles indicate that a
plane of even- and odd-numbered hydrogen atoms lie above and below the plane
of the carbon skeleton, respectively.*

The inelastic neutron TOF spectrum of a butane monolayer adsorbed on Carbopack B at 80 K (10) is shown at the top of Fig. 6. The background inelastic scattering from the substrate has been subtracted. Typical counting time for such a spectrum is ~100 hours. The intramolecular torsional modes of the CH_3 and CH_2 groups observed in the bulk liquid and solid (23) are also found in the monolayer spectrum. In the two methyl torsionsal modes the CH_3 groups rotate either in the same or opposite sense about the terminal C-C bonds, and in the CH_2-CH_2 torsion the two halves of the molecule rotate in the opposite sense about the internal C-C bond. In addition to the intramolecular torsional modes, new features appear in the monolayer spectrum which are not observed in bulk samples: an intense peak at ~112 cm^{-1} and a broader band centered at 50 cm^{-1}.

The appearance of additional peaks in the monolayer spectrum suggests the existence of surface vibratory modes associated with rotations and translations of the free molecule hindered by adsorption. To identify these modes, it is necessary to perform normal mode calculations of the vibrational spectrum of the adsorbed molecule. These calculations are also of interest because of the sensitivity of the frequency and intensity of the surface vibratory modes to the molecular orientation and the location and strength of its bonds to the substrate.

The model calculations (10,19) begin by assuming an orientation of the molecule with respect to the substrate. Interactions between adsorbed molecules are neglected. The molecule-substrate interaction is described by force constants directed along the surface normal, connecting particular atoms in the butane molecule with an infinitely massive substrate. The normal mode problem is then solved assuming only weak coupling to the surface so that the intramolecular force constants and atomic positions of butane are unperturbed. Once the eigenfrequencies and eigenvectors of the vibrational states have been determined, the relative intensity of the modes can be calculated from Eq. (1). As mentioned previously, an advantage of using the graphitized carbon powder substrate is that the dot product in Eq. (1) can be replaced by its random average.

Calculated spectra for different orientations and bonding schemes of the butane molecule are shown in the lower portion of Fig. 6. In Fig. 6(a) the plane of the carbon skeleton is taken to be parallel to the surface (see Fig. 5) and two different force constants are introduced between the surface and the bottom layer of co-planar hydrogen atoms in the molecule. To optimize the fit to the observed spectrum, it was found necessary to use a larger force constant for the CH_2 hydrogen-surface bond (0.065 mdyn/Å) than for the CH_3 hydrogen-surface bond (0.02 mdyn/Å). A second bonding scheme in the same molecular orientation (Fig. 6(b)) introduced equal force constants between each butane carbon atom and the surface but neglected the surface bonding of the hydrogen atoms. Also, a different molecular orientation was tried with the

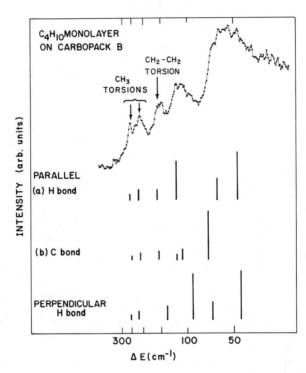

Surface Science

Figure 6.   Comparison of observed and calculated vibrational spectra for a butane monolayer (19). (Top) Observed spectrum for monolayer butane adsorbed on a graphitized carbon powder at ∼ 80 K. The background inelastic scattering from the substrate has been subtracted. (a) Calculated spectrum for the butane molecule adsorbed with its carbon skeleton parallel to the graphite layers and the bottom layer of four hydrogen atoms bonded to the surface with force constants listed in Table I. (b) Same orientation but only the carbon atoms are bonded to the surface with a force constant of 0.12 mdyn/Å. (Bottom) Butane carbon plane perpendicular to the graphite layers and the bottom layer of four hydrogen atoms bonded to the surface with the same force constants as in the parallel orientation.

butane molecule lying on its side but with the plane of the carbon
skeleton perpendicular to the surface. In this orientation there
is again a bottom layer of four co-planar hydrogen atoms--two from
a methyl group and two from the adjacent methylene group. As
before, different force constants were used for the CH_2 and CH_3
hydrogen-surface bonds.

Each of the model spectra in Fig. 6 predicts three surface
vibratory modes: two rocking modes about the principal symmetry
axes of the molecule parallel and perpendicular to the chain
direction, respectively, and a uniform oscillatory motion of the
entire molecule normal to the surface. In the simplified bonding
schemes of these models there is neither a restoring force for a
libration about an axis normal to the surface nor for translational modes parallel to the surface. As can be seen in Fig. 6
both the frequencies and relative intensities of the surface
vibratory modes vary greatly depending on the molecular orientation and bonding to the surface. For example, in the H-bonded
plane-parallel configuration of Fig. 6(a), the rocking mode about
the chain axis occurs at the highest frequency followed by the
bouncing mode normal to the surface and the orthogonal rocking
mode. However, with carbon-surface bonding in this orientation
(Fig. 6(b)), the order of the latter two modes is reversed. These
differences emphasize the need with even moderately sized molecules for performing normal mode calculations in order to identify
the eigenvectors of the surface vibratory bands.

In these model calculations, the best fit to the observed
spectrum was obtained with the butane carbon plane parallel to the
surface with bonds to the bottom layer of hydrogens (Fig. 6(a)).
The agreement worsened when carbon-substrate bonds were included
in this orientation and became quite bad when the hydrogen-
substrate bonds were neglected entirely (Fig. 6(b)). With the
carbon plane perpendicular to the surface (Fig. 6(c)), no combination of force constants yielded as good a fit as in the plane-
parallel configuration.

To what extent, then, can inelastic neutron spectra together
with model calculations of this type be used to infer the orientation and forces bonding a molecule to a surface? It appears
feasible to differentiate between high-symmetry configurations of
an adsorbed molecule--e.g., in the case of butane on a graphite
basal plane, whether the carbon skeleton is parallel or perpendicular to the surface. However, without supporting evidence from
other experiments such as elastic diffraction (see Sec. III), it is
difficult to analyze less symmetric configurations. In these
cases, the parameter space describing the substrate and intermolecular force fields will, in general, be too large to include
the Euler angles of the molecule as free parameters. A quantitative characterization of the molecule-substrate interaction
will also be difficult without independent evidence of the molecular orientation. The calculations based upon empirical potentials discussed in the next section suggest that the force

constants inferred above from the analysis of the vibrational spectra may be accurate to about a factor of three. If the molecular orientation is known, it seems possible to determine at least qualitatively the relative strength of bonding of different parts of the molecule to the substrate.

3. Application of empirical atom-atom potentials to the calculation of the structure and dynamics of butane adsorbed on graphite.

One of the advantages of the hydrocarbon-graphite systems is that they offer the possibility of calculating the molecule-substrate interaction from empirically derived atom-atom potentials. A considerable effort has been made in the last fifteen years in developing empirical potentials of the van der Waals or non-bonding type to describe the interactions between molecules in crystalline hydrocarbons (24,25). Hansen and Taub (26) have studied the structure and dynamics of single paraffin molecules, ethane, propane, and butane, adsorbed on a graphite basal plane using C-C and C-H interatomic potentials of the form

$$E = Ar^{-6} + Bexp(-Cr)$$

where r is the interatomic distance and A, B, and C are parameters characteristic of each atomic pair.

A contour plot of the minimum potential energy of a butane molecule adsorbed on a graphite basal plane is shown in Fig. 7. The potential energy is calculated from a pairwise sum of the interaction of each carbon and hydrogen atom of the molecule with those carbon atoms of the graphite which are within a specified cut-off length. At each point within the basis triangle the molecule is rotated about its three symmetry axes and translated normal to the surface until an energy minimum is found. The most favorable configuration (Fig. 7(c)) corresponds to the molecule lying on its side with the plane of the carbon skeleton parallel to the graphite basal plane. The four co-planar hydrogen atoms nearest the surface (one from each CH_2 and CH_3 group) occupy the centers of the graphite carbon hexagons. This orientation agrees with that deduced above from the analysis of the monolayer vibrational spectrum but differs somewhat from the results of elastic neutron diffraction experiments (Sec. III).

The empirical potentials can also be used to calculate the frequencies of the surface vibratory modes. Two different methods have been employed (26). In the first case, the molecule-substrate force constants introduced in the model described above are calculated and then used to solve the normal mode problem as before. In the second method, the molecule is treated as a rigid body, since distortions of the molecule induced by adsorption are calculated to be small. The frequencies of the surface vibratory modes are computed from the curvature of the molecule-substrate potential as the rigid molecule is rocked about its two symmetry

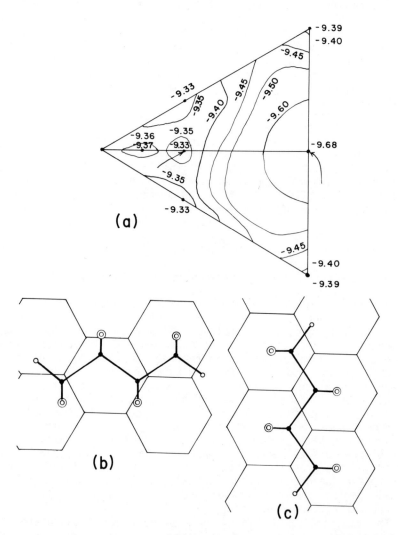

Physical Review B

Figure 7. *(a) Minimum potential-energy contours in kcal/mol for butane on a graphite basal plane (26). At all points in the basis triangle the potential energy is minimized with the butane carbon skeleton parallel to the surface. The arrows mark points for which the molecular configuration is shown: (b) the least favorable configuration,* $E = -9.33$ *kcal/mol; (c) the equilibrium configuration,* $E_{eq} = -9.69$ *kcal/mol.*

axes parallel to the surface and translated along the surface normal.

Table I summarizes some of the results of the dynamical calculations for adsorbed butane. The calculated surface vibratory mode frequencies are in reasonable agreement with the observed spectrum, lying in the range 50-125 cm^{-1} with the rocking mode about the chain axis having the highest frequency followed by the closely spaced bouncing and orthogonal rocking modes. Although there is some variation depending on the force-constant model used, the calculated frequencies are within 30 cm^{-1} of the experimental values.

The empirical potential calculations also give the perturbation of the intramolecular torsional modes of the adsorbed butane. A negligible shift is predicted for the CH_3 torsions while a ~25% increase over the free-molecule frequency is found for the CH_2-CH_2 torsion in the adsorbed molecule. This frequency shift is in reasonable agreement with that observed for the CH_2-CH_2 torsion between the monolayer and bulk liquid spectrum (23) which best approximates that of the free molecule.

The most serious discrepancy between the model fit to the observed spectrum and the calculations based on empirical potentials is that the latter predict the same force constant for each of the hydrogen-surface bonds. The calculated value of 0.02 mdyn/Å is fortuitously close to that obtained for the CH_3 hydrogen-surface bond in the fit to the observed spectrum but about a factor of three smaller than the fitted value for the CH_2 hydrogen-surface bond. The equivalence of the H-surface bonds in the empirical potential calculations is a consequence of each of the hydrogen atoms being situated at the center of a graphite carbon hexagon (cf. Fig. 7(c)). However, even if the hydrogen atoms in the bottom layer of the molecule were not at symmetry-equivalent sites (e.g., out-of-registry could be induced by the butane-butane interaction neglected here), one would not expect a large variation in the H-surface force constant over the graphite basal plane.

The calculations with empirical potentials of the structure and dynamics of paraffin molecules adsorbed on graphite are presently being extended (27) to include the intermolecular interaction. This work together with elastic neutron diffraction experiments on deuterated paraffin films (see Sec. III) should provide a sounder basis for interpreting the inelastic neutron vibrational spectra.

D. Inelastic neutron spectroscopy applied to molecules chemisorbed on catalytic substrates. The model system considered in the previous section is of interest in demonstrating the degree to which neutron vibrational spectra can be interpreted on a well-characterized substrate. Unfortunately from the standpoint of this symposium, the graphite substrates in these experiments are not chemically active. Therefore, in this section, we wish to

TABLE I. Dynamical properties of butane adsorbed on graphite. (a) Calculated and fitted force constants between the molecule and the graphite basal plane. Force constants are calculated from empirical atom-atom potentials (Ref. 26) for the equilibrium configuration of the molecule in which the plane of the carbon skeleton is parallel to the surface. Bonds to the butane carbon atoms (C) and the four co-planar hydrogen atoms nearest the surface (one from each CH_2 and CH_3 group) are considered. H_2 and H_3 denote a CH_2 and CH_3 hydrogen-surface bond, respectively. Fitted force constants are for the butane monolayer vibrational spectrum (Ref. 19) assuming the same molecular orientation but without butane carbon-surface bonds. (b) Calculated and observed frequencies of the surface vibratory modes. $\theta_X(\theta_Y)$ designates a rotational mode about the X-axis (Y-axis) parallel to the surface as in Fig. 5 and T_Z is a translational mode normal to the surface. The valence model uses the calculated force constants in (a); the rigid molecule approximation is described in the text.

(a)	Force Constant (mdyn/Å)	H_2	H_3	C
	calculated	0.021	0.021	0.38
	fitted	0.065	0.02	0.0
(b)	Normal mode frequencies (cm^{-1})	θ_X	T_Z	θ_Y
	valence model	92.0	82.5	72.5
	rigid molecule approximation	123	90	78
	observed	112	~50[a]	~50[a]

[a] The observed T_Z and θ_Y modes of adsorbed butane appear to be unresolved in a broad band centered at 50 cm^{-1}.

review briefly the progress which has been made in the last few years in applying neutron vibrational spectroscopy to molecules chemisorbed on catalysts.

Catalytic substrates are more difficult to use in neutron scattering experiments because of the problems in obtaining both high-area and oxygen-free surfaces. Thus far, ones which have been used include Raney nickel (28-30) which can be prepared in an oxygen free environment and platinum (31,32) and palladium (33) blacks which can be deoxygenated fairly easily by hydrogenation. Most of the work has involved hydrogen as the adsorbate.

The most recent results for hydrogen chemisorbed on Raney nickel by Kelly et al (30) are shown in Fig. 8. The inelastic neutron TOF spectra are taken at 77 K in the energy-loss mode with the background spectrum of the reduced Raney Ni sample after desorption of adsorbed species (Fig. 8(a)) subtracted. Figure 8(b) is the spectrum for a hydrogen monolayer. Features at ~120 meV (960 cm^{-1})and ~140 meV (1120 cm^{-1})are in excellent agreement with previous results of Renouprez et al (28,29). In addition, a peak at 78 meV (624 cm^{-1}) is clearly resolved.

Due in part to the poor characterization of the poly-crystalline substrate, it is difficult to make a definitive assignment of these modes. Since hydrogen adsorbs on a clean nickel surface dissociatively, it is generally believed that the modes can be assigned to various multiply bonded configurations on nickel. The frequencies of the modes are qualitatively, if not quantitatively, similar to those obtained in model calculations (34-36). However, it has been argued (37) for the two higher modes that the data (28,29) do not really allow any distinction to be made between a multiple or single bonded geometry. The two models yield similar intensity patterns which differ from those observed. A stronger case (30) for the 78 meV (624 cm^{-1}) mode resulting from a hydrogen chemisorbed on a four-fold site can be made by comparing studies of hydrogen absorbed in transition metals.

Kelly et al (30) have also investigated the effect of the co-adsorption of 0.25 monolayers of CO with a hydrogen monolayer on Raney Ni. Comparison of Figs. 8(b) and 8(c) indicates that a substantial fraction of adsorbed hydrogen is neither displaced nor vibrationally perturbed by the CO. Since the catalytic methanation reaction is believed to proceed via the hydrogenation of an active carbide-like layer, these authors also obtained a vibrational spectrum for hydrogen on carbon-covered nickel (Fig. 8(d)). In neither the co-adsorption nor carbide-layer experiments was a clear assignment of the vibrations possible. However, in the case of a carbon-covered surface, evidence was cited for hydrogenic deformation and bending modes in the upper frequency range (>800 cm^{-1}), while the lower frequency peaks were associated with Ni-C vibrations.

Inelastic neutron TOF spectra have also been obtained for hydrogen adsorbed on platinum powders (31,32). Both experiments observed a sharp surface vibratory mode of the hydrogen near 400

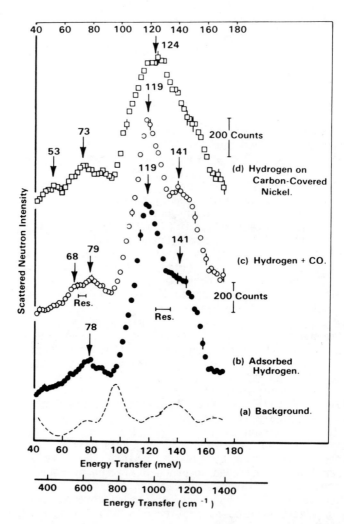

Chemical Physics Letters

Figure 8.   Inelastic neutron spectra from adsorbed species on Raney Ni (30). Each spectrum was smoothed with a 3-point smoothing routine; statistical error bars are indicated at various data points; the lower vertical scale marker refers to Spectra a, b, and c.

cm^{-1} as shown in the spectrum of Fig. 9. This mode is notable for its low frequency compared to that observed for hydrogen adsorbed on nickel (28,30) and palladium (33). It has been suggested that this is a consequence of the high electron density at the surface of platinum leading to multiple bonding of the H atoms (32).

Palladium black is another transition metal substrate on which hydrogen chemisorption has been studied by inelastic neutron scattering (33). Two bands were observed at 916 cm^{-1} and 823 cm^{-1} which were assigned to vibrations of the hydrogen parallel and perpendicular to the surface, respectively. A bridge site was inferred based on the number of modes observed without making any assumptions about the expected excitation frequencies for the different site geometries.

Although not exhaustive, the above summary of experiments with hydrogen chemisorbed on transition-metals serves to illustrate how neutron vibrational spectroscopy is performed with catalytic substrates and the methods used to analyze the inelastic neutron spectra. In concluding this section we note that the technique can be extended to supported catalysts such as in recent experiments with hydrogen adsorbed on both MoS$_2$ and alumina supported MoS$_2$ (38). Also, as another indication of the variety of systems which can be studied, we note earlier experiments with ethylene (39) and acetylene (40) adsorbed on silver exchanged 13X zeolites. In this work, deuteration of the molecules was helpful in identifying the surface vibratory modes on these ionic substrates of greater complexity.

III. Determination of the Orientation of Adsorbed Molecules by Elastic Neutron Diffraction

Elastic neutron diffraction has developed rapidly in the past few years as a method of studying the structure of gases physisorbed on high-surface-area substrates. As discussed in previous reviews, (1-3) the technique has been used primarily to investigate the translational order of adsorbed layers and to elucidate the phase diagrams of simple gases physisorbed on various exfoliated graphite substrates. Our purpose here is to focus on some very recent applications of elastic neutron diffraction to the determination of orientational order in molecular monolayers. We shall return to our model system of butane on graphite as an example of a large class of deuterated hydrocarbon films whose orientational order can be studied by this technique. The results will also be relevant to the interpretation of the monolayer butane vibrational spectrum and the structure calculations with empirical potentials presented in the previous section.

Detailed knowledge of the orientation of an adsorbed molecule is difficult to obtain by conventional surface scattering techniques such as LEED. The reason is that the electron is such a strongly interacting probe that multiple scattering effects greatly complicate the interpretation of the intensity of the Bragg spots from the overlayer. As discussed in Sec. I, the neutron-nuclear interaction is much weaker so that, for most adsorbed systems, it is possible to design samples for which multiple scattering effects are negligible. Indeed, with elastic neutron diffraction the problem has been rather one of obtaining sufficient scattered intensities to observe enough Bragg reflections from a monolayer to determine unambiguously the 2D lattice. Even in the most favorable circumstances such as with ^{36}Ar adsorbed on graphite, (9) only a few Bragg reflections have been observed. The ^{36}Ar diffraction pattern contained three Bragg reflections which were sufficient to confirm a triangular lattice characteristic of most of the simplest gases adsorbed on graphite.

Molecular orientational order in adsorbed monolayers can be inferred indirectly from elastic neutron diffraction experiments if it results in a structural phase transition which alters the translational symmetry of the 2D lattice. In such cases, Bragg reflections appear which are not present in the orientationally disordered state. Experiments of this type have inferred orientational order in monolayers of oxygen (41) and nitrogen (42) adsorbed on graphite. However, these experiments have not observed a sufficient number of Bragg reflections to determine the molecular orientation by comparing relative Bragg peak intensities with a model structure factor.

Recently, it has begun to be appreciated that films of larger molecules which form non-triangular lattices of lower symmetry will exhibit a larger number of Bragg reflections in the Q-range accessible to thermal neutron scattering. In these experiments a conventional structure analysis can be performed in which the relative intensity of the Bragg reflections is interpreted in terms of a geometrical structure factor to yield orientational parameters of the adsorbed molecules. The first experiment of this kind was performed by Suzanne et al (43) with nitric oxide adsorbed on exfoliated graphite. They obtained the diffraction pattern shown in Fig. 10 for a coverage of 0.4 layers of nitric oxide adsorbed on Papyex at 10 K. The Bragg peaks could be indexed by the close-packed unit cell shown in the inset. They assumed that the rectangularly-shaped N_2O_2 dimer was lying down with its plane parallel to the surface, since the molecular area in this orientation is very close to that of the unit cell. The one orientational parameter in their model was the angle betwen the long axis of the molecule and a cell axis. This angle was adjusted to give the best fit to the relative intensity of the five Bragg peaks observed as indicated by the height of the arrows in Fig. 10. A similar structure analysis was performed for the nitric oxide film at higher coverage (1.05 layers) where it was assumed that the N_2O_2 dimers were standing on end, and the adjustable orientational parameter was the angle between the molecular plane and a cell axis.

Deuterated hydrocarbon films have been found to be excep-

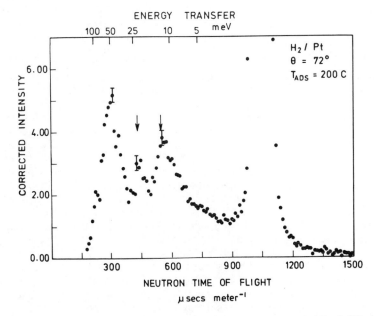

ENERGY   TRANSFER

*Figure 9.   Corrected inelastic neutron spectrum for hydrogen adsorbed on platinum
black at 200°C (32)*

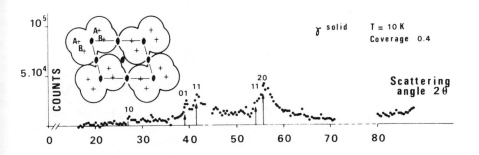

*Figure 10.   Elastic neutron diffraction pattern from a 0.4-layer nitric oxide film
adsorbed on a recompressed graphite powder Papyex (43). Background scattering
from the substrate has been subtracted. The molecules are assumed to be lying
down as shown in the unit cell (inset). The arrows indicate the position and relative
intensities of the Bragg peaks calculated for this structure.*

tionally favorable for study by elastic neutron diffraction because of the large number of Bragg reflections observable at monolayer coverages. Reflections out to third order have been observed in the diffraction patterns of neopentane (C_5D_{12}) adsorbed on TiO_2 (44) and ethane (C_2D_6) (45) and butane (C_4D_{10}) (46) adsorbed on exfoliated graphite. [6] Besides negligible multiple scattering, neutron diffraction has a further advantage over LEED for studies of hydrocarbon overlayers. The coherent neutron cross-section of the film can be greatly enhanced by deuterating the molecules. Because the deuterium and carbon nuclei are equally effective in scattering the neutrons, the relative intensity of the Bragg reflections is sensitive to the position of the deuterium atoms. In contrast, the diffraction from hydrocarbons in LEED experiments is dominated by scattering from the carbon atoms so that there is much less sensitivity to the hydrogen atom positions. Note also that for deuterated hydrocarbons on graphite it is not a large ratio of film-to-substrate scattering which is responsible for the rich neutron diffraction patterns, since the coherent cross-section of the deuterium and carbon are nearly equal. Instead, it is the symmetry of the 2D lattice and the particular nature of the structure factors which results in a large number of Bragg reflections being observed.

In the case of the ethane (45) and butane (46) monolayers adsorbed on graphite, it has been possible to analyze the neutron diffraction patterns using all three Euler angles of the molecule as orientational parameters. Here we limit discussion to the butane monolayer which we have taken as a model system and whose vibrational spectrum was discussed in Sec. II.

Elastic diffraction patterns from a 0.8-layer film of butane on Grafoil are shown for two different neutron wavelengths, λ = 4.07 Å in Fig. 11(a) and 1.10 Å, in Fig. 12. The large momentum transfers obtainable at the shorter wavelength permit seven Bragg reflections to be reached, while the longer wavelength provides improved Q resolution for observing the most intense peaks. In the latter case, the asymmetric shape of the peaks characteristic of diffraction from 2D polycrystals (47) is clearly evident. The unit cell of minimum area found to index the Bragg reflections is shown in Fig. 11(b). The structure is close-packed with one molecule per cell.

The elastic neutron diffraction pattern cannot be interpreted to give the bonding site of the molecule other than what can be inferred indirectly from the size and shape of the monolayer unit cell relative to that of the adsorbing surface. The unit cell in Fig. 11(b) is compatible with partial registry on the graphite basal plane as suggested by the d-spacing of 4.26 Å (three times the length of a carbon hexagon edge). However, the other primitive translation vector of the cell does not permit complete registry in which the hydrogen atoms nearest the surface occupy the centers of the carbon hexagons as predicted by the calculations with empirical potentials (see Fig. 7(c)). Apparently, a competition

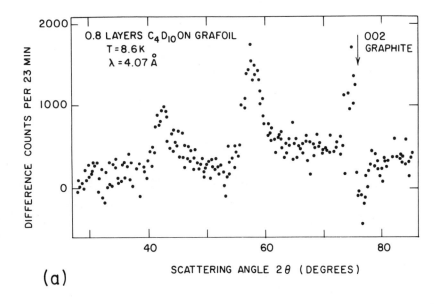

(a)

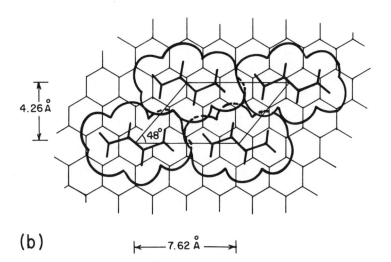

(b)

Figure 11. (a) Elastic neutron diffraction pattern from a 0.8-layer butane film adsorbed on Grafoil at 8.6 K (46). The background scattering from the substrate has been subtracted. Neutron wavelength is 4.07 Å. (b) Unit cell of minimum area that indexes the observed Bragg peaks in (a) (see also Figure 12).

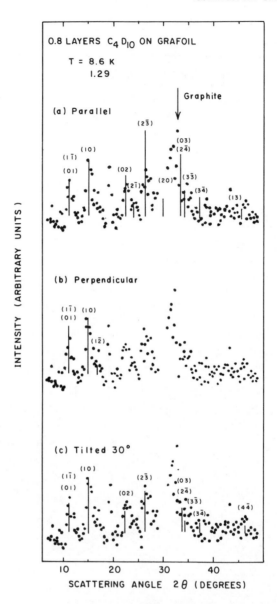

Figure 12.   Comparison of elastic diffraction of butane monolayer as in Figure 11 with calculated spectra for different orientations of the molecule (46). Neutron wavelength is now 1.10Å. Arrow labeled "Graphite" marks region of strong substrate scattering, resulting in large uncertainty in the difference spectrum. The calculated spectra are for the butane molecule adsorbed with its carbon skeleton: (a) parallel to the graphite layers; (b) perpendicular to the layers; and (c) tilted at 30° to the graphite basal plane about the x-axis (see Figure 5). The calculated structure factor differentiates clearly between the parallel and perpendicular orientations. The tilted configuration (c) appears to give the best fit.

between the intermolecular interaction neglected in these calculations and the substrate contribution to the potential energy
produces this partially registered state.

The orientation of the butane molecule was determined by
systematically varying its three Euler angles to find the best
agreement between the calculated and observed Bragg intensities.
To limit the number of fitting parameters, the Debye-Waller factor
was set equal to unity. Of particular interest are the two
orientations considered in the analysis of the butane vibrational
spectrum, i.e., with the chain axis parallel to the graphite basal
planes and with the butane carbon skeleton either parallel or
perpendicular to the surface. The observed diffraction pattern is
compared with the calculated structure factor for these two orientations in Fig. 12. The fit for the perpendicular orientation (b)
is quite poor with only two Bragg peaks predicted to have appreciable intensity. In the parallel configuration (a) the fit is
qualitatively better in agreement with the analysis of the vibrational spectrum in Fig. 6.

Surprisingly, the fit to the observed diffraction pattern can
be further improved by tilting the butane carbon plane at an angle
of 30° to the surface Fig. 12(c). This orientation conflicts with
the plane-parallel configuration predicted by the calculations
with empirical potentials (26). In an effort to understand this
discrepancy, the effect of the butane-butane interaction and electrostatic forces (48), both of which were neglected in the previous
calculations, are now being investigated (27). (The electrostatic
forces arise from charge transfer within the molecule). Also of
concern is whether the atom-atom potentials adequately characterize the semimetallic character of the graphite substrate.
The elastic neutron diffraction experiments may well provide the
most sensitive test of the application of empirical potentials to
the adsorption of hydrocarbons on graphite.

The tilted configuration of butane was not considered in the
analysis of the monolayer vibrational spectrum in Sec. II.C.2.
Normal mode calculations are now being performed (20) with this
orientation to see if the fit to the observed spectrum can be
improved. The tilting of the molecule may be related to the
inconsistency encountered in the plane-parallel model in which
different atom-substrate force constants had to be introduced for
the co-planar CH_2 and CH_3 hydrogens. In the tilted configuration
this bottom layer of hydrogens is split into two separate levels of
atoms. This difference in height above the surface may provide a
physical basis for two different force constants.

IV. Conclusions and Prospects for the Future

We have seen that inelastic neutron scattering at steady
state reactors provides a spectroscopic technique for studying the
dynamics of adsorbed molecules over a wide energy range, $\sim 10^{-5}$ V to

$\sim 10^{-1}$ V. With the development of pulsed spallation neutron sources now under construction in the U.S. and Europe (49)[1] significantly higher neutron fluxes at energies ~ 0.5 V (4000 $\overline{cm}^{-1}$) will become available in the next few years. Thus a spectral range which up to now has been dominated by electron and optical spectroscopies will soon become available to neutron scattering with its unique features as a surface probe. Foremost among these for catalytic systems are the penetrability of the neutron combined with its sensitivity to hydrogenous species. The neutron penetrability enables experiments on adsorbed systems at high ambient pressures, suggesting a significant role for neutron scattering in bridging the "pressure gap" between classical thermodynamics techniques and modern high-vacuum surface spectroscopies.

In this paper we have considered a model system in order to assess under the most favorable conditions the extent to which neutron spectroscopy can be used to infer the orientation of an adsorbed molecule and the forces bonding it to a substrate. Although still in an early stage of development, elastic neutron diffraction offers promise of being a powerful technique for determining molecular orientation in deuterated hydrocarbon monolayers. Once the orientation of the adsorbed molecule is known, the interpretation of the vibrational spectrum can be made much more compelling. Here the advantage of neutron spectroscopy is knowledge of the scattering laws, allowing quantitative comparison of the observed frequencies and intensities of the surface vibratory modes with model calculations.

Perhaps the biggest question in the future of neutron spectroscopy is the the extent to which the progress with graphite substrates can be carried over to catalytic systems. Other high-area powders are needed with uniform surfaces and in which a single crystal plane provides the majority of adsorption sites. In this respect, some success has already been reported for metal-oxide powders such as MgO (50) and TiO$_2$ (44) and for some high-area metal powders (51). Possibly these substrates can provide an intermediate step between the graphites and commercial catalysts. Also, it may be worthwhile investigating the vibrational states of larger molecules than hydrogen on presently available transition-metal powders. As we have seen, an alkane such as butane can exhibit intense surface vibratory modes which are sensitive to the local environment of the molecule. But ultimately, substrate quality would appear to be the decisive factor in determining future contributions of neutron scattering to surface spectroscopy.

IV. Acknowledgement

The author wishes to acknowledge partial support of this work by the National Science Foundation under Grant No. DMR-7905958 and the Petroleum Research Fund administered by the American Chemical Society.

Abstract

This paper treats neutron scattering spectroscopy of adsorbed molecules within the framework of two basic questions: the extent to which the technique can be used to determine, first, the orientation and position of an adsorbed molecule; and, second, the strength and location of forces bonding a molecule to a surface. After considering the distinguishing features of the neutron as a surface probe, the applications of inelastic neutron scattering to the vibrational spectroscopy of adsorbed molecules and elastic neutron diffraction to the study of film structures are briefly reviewed. A model system consisting of monolayer n-butane ($CH_3(CH_2)_2CH_3$) adsorbed on a powdered graphite substrate is used to illustrate the capabilities and limitations of the technique. Several aspects of this system are discussed: 1) the modeling of the vibrational spectra observed by inelastic neutron scattering to infer the orientation of the adsorbed molecule and the force constants characterizing its interaction with the substrate; 2) the determination of the two-dimensional unit cell and the molecular orientation from analysis of the elastic neutron diffraction pattern of the monolayer; and 3) calculation of the film structure and dynamics from empirically derived interatomic potentials. Discussion of the model system is followed by a short summary of recent work by several investigators on the dynamics of hydrogen chemisorbed on transition-metal powders. The paper concludes with prospects for the future based on both the model system results and experiments with transition-metal catalysts.

Literature Cited

1. White, J.W.; Thomas, R.K.; Trewern, I.; Marlow, I.; Bomchil, G. Surf. Sci., 1978, 76, 13.

2. Hall, P.G.; Wright, C.J. in "Chemical Physics of Solids and Their Surfaces", Robert, M.W.; Thomas, J.M., Eds.; Chemical Society: London, 1978; Vol. 7, p. 89.

3. McTague, J.P.; Nielsen, M.; Passell, L.; Crit. Rev. Solid St. Sci., 1979, 8, 135.

4. See, for example, Marshall, W.; Lovesey, S.W. "Theory of Thermal Neutron Scattering"; Oxford University Press: Oxford, 1971.

5. See also, Bacon, G.E. "Thermal Neutron Diffraction", 3rd ed.; Oxford University Press: Oxford, 1976, for a discussion of spin-incoherent scattering as well as other aspects of the theory and practice of neutron scattering.

6. Gurevich, I.I.; Tarasov, L.V. "Low Energy Neutron Physics"; North-Holland: Amsterdam, 1968.

7. Mills, D.L.; Surf. Sci., 1975, 48, 59.

8. Pearce, H.A.; Sheppard, N. Surf. Sci., 1976, 59, 205.

9. Taub, H.; Carneiro, K.; Kjems, J.K.; Passell, L; McTague, J.P. Phys. Rev. B, 1977, 39, 215.

10. Taub, H.; Danner, H.R.; Sharma, Y.P.; McMurry, H.L.; Brugger, R.M. Phys. Rev. Letters, 1977, 39, 215.

11. Boutin, H.; Prask, H.; Iyengar, R.D. Adv. in Col. and Inter-face Science, Overbeek, J.T.G.; Prins, W.; Zettlemoyer, A.C., Eds.; 1968, 2, (1).

12. Verdan, G. Phys. Letters, 1967, 25A, 435.

13. Todireanu, S.; Hautecler, S. Phys. Letters, 1973, 43A, 189.

14. Todireanu, S. Nuovo Cimento Suppl., 1967, 5, 543.

15. Nielsen, J.; McTague, J.P.; Ellenson, W.D. J. Phys. (Paris), 1977, C-4, 10.

16. Silvera, I.F.; Nielsen, M. Phys. Rev. Letters, 1976, 37, 1275.

17. Newbery, M.W.; Rayment, T.; Smalley, M.V.; Thomas, R.K.; White, J.W. Chem. Phys. Letters, 1978, 59, 461.

18. For studies of the translational diffusion of methane on graphite see White, J.W.; Thomas, R.K.; Gamlen, P.H. "NBRC Annual Report"; SRC: London, 1976 and Coulomb, J.P.; Bienfait, M.; Thorel, P. Phys. Rev. Letters, 1979, 42, 733.

19. Taub. H.; Danner, H.R.; Sharma, Y.P.; McMurry, H.L.; Brugger, R.M. Surf. Sci., 1978, 76, 50.

20. Wang, R.; Danner, H.R.; Taub, H., to be published.

21. Stong, K.A.; Brugger, R.M. J. Chem. Phys., 1967, 47, 421.

22. Brugger, R.M.; Strong, K.A.; Grant, D.M. in "Neutron Inelastic Scattering"; IAEA: Vienna, 1968; Vol. II, p. 323.

23. Strong, K.A. "Catalogue of Neutron Molecular Spectra"; AEC Report IN-1237, Idaho Nuclear Corp.

24. Williams, D.E. J. Chem. Phys., 1966, 45, 3770; 1967, 47, 4680.

25. Kitaigorodskii, A.E. "Molecular Crystals"; Academic: New York, 1973.

26. Hansen, F.Y.; Taub, H. Phys. Rev. B, 1979, 19, 6542.

27. Hansen, F.Y.; Taub, H., to be published.

28. Renouprez, A.J.; Fouilloux, P.; Coudurier, G.; Tochetti, D.; Stockmeyer, R. J. Chem. Soc. Faraday I, 1977, 73, 1.

29. Stockmeyer, R.; Conrad, H.M.; Renouprez, A.; Fouilloux, P. Surf. Sci., 1975, 49, 549.

30. Kelly, R.D.; Rush, J.J.; Madey, T.E. Chem. Phys. Letters, 1979, 66, 159.

31. Asada, H.; Toya, T.; Motahashi, H.; Sakamoto, M.; Hamaguchi, Y. J. Chem. Phys., 1975, 63, 4078.

32. Howard, J.; Waddington, T.C.; Wright, C.J. J. Chem. Phys., 1976, 64, 3897; "Neutron Inelastic Scattering"; IAEA: Vienna, 1978; Vol. II, p. 499.

33. Howard. J.; Waddington, T.C.; Wright, C.J. Chem. Phys. Letters, 1978, 56, 258.

34. Andersson, S. Chem. Phys. Letters, 1978, 55, 185.

35. Upton, T.H.; Goddard, W.A. Phys. Rev. Letters, 1979, 42, 472.

36. Melius, C.F.; Upton, T.H.; Goddard. W.A. Solid State Commun., 1978, 28, 501.

37. Wright, C.J. J. Chem. Soc. Faraday II, 1977, 73, 1497.

38. Wright C.J.; Fraser, D.; Moyes, R.B.; Riekel, C.; Sampson, C.; Wells, P.B., preprint, July, 1979.

39. Howard, J.; Waddington, T.C.; Wright, C.J. J. Chem. Soc. Faraday II, 1977, 73, 1768.

40. Howard, J.; Waddington, T.C. Surf. Sci., 1977, 68, 86.

41. Nielsen, M.; McTague, J.P. Phys. Rev. B, 1979, 19, 3096.

42. Eckert, J.; Ellenson, W.D.; Hastings, J.B.; Passell, L. Phys. Rev. Letters, 1979, 43, 1329.

43. Suzanne, J.; Coulomb, J.P.; Bienfait, M.; Maticki, M.; Thomy, A.; Croset, B.; Marti, C. Phys. Rev. Letters, 1978, 41, 760.

44. Marlow, I. Part II Thesis; Oxford University, 1977 as quoted in Ref. 1.

45. Coulomb, J.P.; Biberian, J.P.; Suzanne, J.; Thomy, A.; Trott, G.J.; Taub, H.; Danner, H.R.; Hansen, F.Y. Phys. Rev. Letters, 1979, 43, 1878.

46. Trott, G.J.; Taub, H.; Hansen, F.Y.; Danner, H.R., to be published.

47. Warren, B.E. Phys. Rev., 1941, 59, 693.

48. Williams, D.E.; Starr, T.L. Acta Cryst., 1977, A33, 771.

49. See, for example, Carpenter, J.M.; Blewitt, T.H.; Price, D.L; Werner, S.A. Phys. Today, 1979, 32, (12), 42-49.

50. Dash, J.G.; Ecke, R.; Stoltenberg, J.; Vilches, O.E.; Whittemore, Jr., O.J. J. Phys. Chem., 1978, 82, 1450.

51. Genot, B., Thesis, University of Nancy, 1974 as quoted in Ref. 50.

RECEIVED June 3, 1980.

INDEX